Errors in Practical Measurement in Surveying, Engineering, and Technology

Errors in Practical Measurement in Surveying, Engineering, and Technology

B. Austin Barry, F.S.C., P.E.
Professor of Civil Engineering
Manhattan College, New York

Edited by M.D. Morris, P.E.

Landmark Enterprises
10324 Newton Way
Rancho Cordova, CA 95670

This book was originally published in 1978 by John Wiley and Sons, Inc.
The Library of Congress Cataloging in Publication Data for that edition was:

Barry, Benjamin Austin
Errors in practical measurement in science, engineering, and technology

"A Wiley-Interscience publication."
Bibliography: p.
Includes Index
1. Mensuration. 2. Errors, Theory of . I. Morris, Morton Dan. II. Title

T50.B317 602'.8 78-9751
ISBN 0-471-03156-9
Printed in the United State of America

This edition is printed in the United States of America by:

Data Reproductions
Rochester Hills, Michigan

ISBN: 0-910845-47-6

Permissions Department
Landmark Enterprises
10324 Newton Way
Rancho Cordova CA 95670-2096

To those who guided me
in the error of my ways.

Preface

This is a book about statistical measurement and error theory. Error is not new, but getting a handle on it, and on its tie-ins with probability and with certitude, is. Through a knowledge of variations in measured results, the behavior of errors can be understood; thus a grasp can be had of where correct value lies.

This book covers as painlessly as possible the elements of statistical error analysis. Those whose interest is kindled may go the whole statistics route: books in this area are both profound and abundant. Those who elect not to will at least have examined the range of practical operational means to certainty in measurement.

Why measure something more than once? What can a repeated measurement tell about the true value of the quantity being measured? When is a measurement precise? The notion of variation in measured results leads to the notion of error, and this comes 'round full circle to accuracy and its clue to certitude. And when we believe we have accurate measurement in our grasp, we can confidently expect to achieve nearness to the true value.

That's what this book is all about. People ought to be able to read along with even a bit of relaxed enjoyment to master this not too difficult segment of new knowledge.

BRO. B. AUSTIN BARRY

New York, New York
Fall 1978

Contents

1 INTRODUCTION TO MEASUREMENT, 1

1.1 Engineering and Scientific Measurements, 1
1.2 How Measurements Are Used, 1
1.3 Comparison with Standard, 2
1.4 Standard of Length, 2
1.5 Confusion of Length, 3
1.6 The Official Inch of 1960, 4
1.7 Availability of Standards, 4
1.8 Direct and Indirect Measurements, 6

2 MEASUREMENT ERRORS, 7

2.1 Readings, 7
2.2 Repeated Readings, 8
2.3 Best Value, 9
2.4 Mistakes, 9
2.5 Discrepancy, 9
2.6 Systematic Errors, 10
2.7 Types of Systematic Error, 10

2.7.1 *Natural Errors*, 11
2.7.2 *Instrumental Errors*, 11
2.7.3 *Personal Errors*, 11

2.8 Counteracting Systematic Errors, 11
2.9 Detecting Systematic Errors, 12
2.10 Accidental Errors, 13
2.11 Discrepancies Indicative of Accidental Errors, 13
2.12 Use of the Mean, 14

3 RELIABILITY OF MEASUREMENTS, 15

3.1 Accuracy Versus Precision, 15

3.1.2 *Accuracy*, 15

3.2 Results versus Method, 16
3.3 Behavior and Occurrence of Accidental Errors, 17
3.4 Frequency-Density Curve, 18
3.5 Normal Distribution Curve, 19
3.6 Confidence Inspired by Normal Distribution, 21
3.7 Skewed Frequency-Density Curve, 22
3.8 Least Squares Principle, 22
3.9 Errors and Residuals, 23
3.10 Equating Residuals with Accidental Errors, 25
3.11 Precision an Indicator of Accuracy, 26

4 PROBABILITY THEORY OF ERRORS, 27

4.1 Mean Square Error, 27
4.2 Standard Deviation, 27
4.3 Standard Deviation of the Population, 29
4.4 Number of Values in a Sample or Set, 29
4.5 Standard Error, 30
4.6 Meaning of the Standard Error, 31
4.7 Use of the Standard Error, 31
4.8 Frequency Distribution Table, 32
4.9 Short Method for Standard Deviation, 34
4.10 Use of Standard Errors, 39
4.11 Plotting the Histogram and the Frequency-Density Curve, 39
4.12 Normal Probability Distribution Curve, 45

5 MEASURES OF RELIABILITY, 46

5.1 Definitions, 46
5.2 Mean, 46
5.3 Median, 46
5.4 Mode, 47
5.5 Scatter, 48
5.6 Range, 48
5.7 Mean Deviation, 48
5.8 Variance and Standard Deviation, 49
5.9 The Sigma Error, 50
5.10 The Two-Sigma Error, 51
5.11 The Three-Sigma Error, 52
5.12 Probable Error, 52
5.13 Meaning of Standard Deviation and Standard Error, 53
5.14 Probability Applications, 59

6 RELIABILITY OF REPEATED MEASUREMENTS, 66

6.1 Multiple Measurements, 66
6.2 Number of Repeated Measurements, 67
6.3 Rejecting Measurements, 68
6.4 Maximum Possible Error, 68
6.5 Use of Maximum Error, 69
6.6 Significance, 69
6.7 Using the Standard Error to Compare Sets, 69
6.8 Standard Error of the Difference Between Means, 70
6.9 Limitation of Standard Error of the Difference, 71

7 PROPAGATION OF ERRORS IN COMPUTING, 72

7.1 Introduction, 72
7.2 Addition of Values Containing Errors, 72
7.3 Subtraction of Values Containing Errors, 73
7.4 Multiplication of Values Containing Errors, 74

7.5 Division of Values Containing Errors, 75
7.6 Multiplication of Values Containing Errors by a Constant, 76
7.7 Elevation to a Power of a Quantity Containing an Error, 76
7.8 Root of a Quantity Containing an Error, 78
7.9 Other Operations, 78

8 ERRORS AND WEIGHTS, 80

8.1 Weight and Reliability, 80
8.2 The Weighted Mean, 80
8.3 Assigning Weights, 81
8.4 Equal Weights, 84
8.5 Equal Weights, with Exact Check Value, 84
8.6 Equal Weights, with Arbitrary Check Value, 85
8.7 Equal Weights of Components and Total, 86
8.8 Weights Proportional to Number of Measurements, 87
8.9 Weights and Errors, 87
8.10 Weights and the Standard Error, 87
8.11 Weights and Adjustments, 88
8.12 Unequal Weights, with Exact Check Value, 89
8.13 Standard Error of the Mean of Several Means, 90
8.14 Illustrative Calculations, 91

9 PRACTICAL APPLICATION OF THE THEORY OF ERRORS IN MEASUREMENT, 94

9.1 Remarks, 94
9.2 Standard Deviation a Criterion, 94
9.3 Fixing on a Maximum Desired Error, 95
9.4 Selecting a "Maximum Error", 95
9.5 Procedure for Limiting the Error, 96
9.6 Standardizing the Procedure, 96
9.7 Utilizing a Standard Procedure, 97
9.8 Setting Specifications for a Standard Procedure, 100
9.9 Selecting the Number of Measurements To Be Made, 101

10 TWO-DIMENSIONAL ERRORS, 103

10.1 General, 103
10.2 Definition, 103
10.3 The Probability Ellipse, 104
10.4 The Probability Circle, 105
10.5 Elliptical (Circular) Error Evaluation, 107
10.6 Application to Position Accuracy, 109
10.7 Use of Control Systems, 111

BIBLIOGRAPHY, 112

APPENDICES

A. Significant Figures in Measurement, 113
B. Basic Concepts of Probability and the Normal Probability Curve, 118
C. Writing a Sample Specification for a Procedure, 130
D. Classification, Standards of Accuracy, and General Specifications of Geodetic Control Surveys (National Ocean Survey, NOAA—Feb, 135
E. The Geoid, 144
F. The Frequency Distribution Curve, 151
G. Computer and Calculator Solution of Problems, 159
H. Problems, 165

INDEX, 181

Errors in Practical Measurement in Science, Engineering, and Technology

One

Introduction to Measurement

1.1 Engineering and Scientific Measurements

Basic to all engineering is design, and basic to all design is the making of measurements. In both science and engineering, the collection of information means measurements. Once measurements have been made, they must be organized, evaluated, and interpreted.

Whenever measurements are made, errors are made—the single exception being when the measurement is a discrete count (e.g., the number of people in a room). Because no measurement is free from error, steps must be taken to evaluate the accuracy and the precision of the measurement. To preclude a false sense of accuracy, one must investigate the nature of error, as well as the sources, types, and magnitude of errors made at various stages of the measurement operation, and the interrelation among errors. Only thus is it possible to predict the order of magnitude of the error in the final result.

1.2 How Measurements Are Used

This treatise deals with linear measurements, as with scale, rod, tape, or light waves, and angular measurements, as with protractor, transit, theodolite, or gyroscope. Errors (personal, instrumental, and natural) are studied. They are analyzed to determine whether they are cumulative or in some measure compensatory. Accuracy and precision are differentiated, and analysis of measurement is made with a view to designing systems and specifications to achieve certain desired results. Rules of thumb and empirical formulas are analyzed to determine their worth to the engineer, and modern-day tolerances are also examined.

The use of measurements is the lot of every engineer, and most engineers spend some of their time planning and designing with measured quantities. There are few engineers who do not themselves make measurements. Although it is true that many of these measurements are made by technicians and subprofessionals, their planning and direction is the full responsibility of the engineering professional, whose understanding of the basic principles must guide the planning, design, and execution of the project.

MEASUREMENTS AND STANDARDS

1.3 Comparison with Standard

The word *measurement* implies comparison of a quantity with a standard value of some sort. The quantity to be assessed (length, weight, direction, time, volume, etc.) is measured directly or indirectly against a standard. Throughout history we have had standards, more or less fixed: the cubit (elbow to fingertip), the foot (just that), the inch (thumb joint to tip), and so on. Later, efforts to achieve standards of length shifted to marching men: the pace, the stadia (stride of a man), the mile (1000 strides), and so on. It is said that early map makers had a difficult job because standards varied from place to place, and reported distances when plotted for map construction resulted in some weird configurations. The Romans, too, injected confusion purposely as a defense measure, to keep their many enemies from learning true distances.

A later standard, the meter, was intended to be one ten-millionth of the distance from equator to pole on the earth's surface, although subsequent observations show that it missed this by a little bit. The inch and the yard (both British) were independently developed, and are related to the meter today through official (and legal) conversion factors.

1.4 Standards of Length

The word *inch*, which is derived from the Latin *uncia*, or twelfth part, has been used to measure things in the Anglo-Saxon world since before

William the Conqueror. In those days an inch was defined as the width of a man's thumb, an inadequate standard at best. The first real effort to put the inch on a precise basis was made in the days of Edward I (1272–1307), when it was defined as equal to three barley corns, dry and round, laid end to end. This standard proved to be fairly accurate and was used for several centuries. It is only in the past 200 years that real progress has been made in the field of calibration.

In 1866 the United States for the first time linked the inch to an international standard when, by act of Congress, it pronounced the international prototype meter at Paris as equal to 39.37 in. A few years later, as a result of its participation in the Metric Convention of 1875, this country received a platinum-iridium meter, which became the official standard for all American measures of length. In the ensuing decades, the calibration was made even more accurate, and by 1940 the National Bureau of Standards (NBS) was able to define an inch accurate to one-millionth of an inch. But it was not allowed to rest on this notable accomplishment. A few years ago a group of expert metrologists from the machine-tool industry informed the NBS that the accuracy of existing gauge blocks was no longer sufficient to meet their advanced needs. They urged the development of an inch accurate to one or two ten-millionths of an inch, a project that is being actively pursued by the Bureau with the collaboration of private industry.

1.5 Confusion of Length

But, as many experts pointed out, refinement of the American inch properly called for standardization with the British and Canadian inches. Because of the 1866 act of Congress, the American inch was equal to 2.540005 cm. The British inch was equal to 2.539995 cm, and most experts wanted the United States to adopt the British standard. That infinitesimal difference of 0.00001 cm had immensely complicated the manufacture and interchangeability of precision instruments on either side of the Atlantic during World War II.

But the proposal to adopt the British inch ran into opposition. Although it was backed by the Army, the Navy, the National Advisory Committee for Aeronautics, and the NBS, the Coast and Geodetic Survey (now the National Geodetic Survey) pointed out that the changeover would present it with formidable problems. The Survey had established plane coordinate systems for each of 48 states about 30 years before. Subsequently, similar systems were established for Hawaii and Alaska. These map systems include some 150,000 triangulation and traverse points, each of which would have had to be changed by several feet if the British standard had been adopted. By mutual consent, the special problems of the Coast and Geodetic Survey were recognized, and it continued to use the old-style inch. Most other official bodies of the United States government, however, are switching over to the new inch (equal to 2.54 cm exactly), as the National Advisory Committee on Aeronautics has already done for use in altimetry and airspeed computations and in defining the standard atmosphere.

With the inch set officially at 2.54 cm exactly, the problem has shifted to the definition of the standard meter. Experience has shown that even bars of platinum-iridium kept hermetically sealed at constant temperature are not changeless. The process of atomic disintegration changes them slightly, decade by decade.

1.6 The Official Inch of 1960

But the scientists of the world worked busily on an atomic standard for length measurement. In 1960 international agreement was reached on a specific atomic wavelength, the so-called orange line of the isotope krypton 86 (^{86}Kr). The inch was defined as 2.54 cm exactly, and the meter as "equal to 1,650,763.73 wavelengths in vacuum of the radiation corresponding to the transition between the energy levels of 2_{p10} and 5_{d5} of the atom Krypton 86."

The inch, one might have thought, would now be pretty well nailed down, at least for the foreseeable future. In 1970, however, the NBS announced its development of a new length standard using the methane/stabilized helium-neon (He-Ne) laser. Reproducible to within a part in 10^{11}, this laser is a likely candidate to succeed ^{86}Kr as the standard of length.

1.7 Availability of Standards

Since time, temperature, voltage, weight, angular, gravity, and frequency standards have been developed, it is aptly remarked that we live today in a standardized world. The *ampere* is defined as the magnitude of the current that when flowing through each of two long parallel wires separated by one meter in free space, results in a force between the two wires (due to their magnetic field) of 2×10^{-7} newton for each meter of length. The thermodynamic or Kelvin scale of *temperature* used in the International System (SI) has its origin or zero point at absolute zero and has a fixed point at the triple point of water, defined as 273.16°K. The *candela* is defined as the luminous intensity of 1/600,000 of a square meter of a radiating blackbody at the temperature of freezing platinum (2042°K).

An absolute determination of *gravity* made at Potsdam in 1904 became, by international agreement, the standard to which all other gravity measurements were referred. The NBS has made a new absolute determination (1967) of the acceleration due to gravity (g), the value at Gaithersburg, Maryland, being 9.801018 m/s^2 with a standard deviation of $\pm 0.3 \times 10^{-5} m/s^2 (= \pm 0.3$ milligal).

Time standards, traditionally maintained at celestial observatories on "standard" clocks checked by star sights, have recently been geared to an invariant "atomic clock." The *second* of time is currently defined as the duration of 9,192,631,770 cycles of the radiation associated with a specified transition of the cesium atom. The NBS time signals broadcast on WWV in Colorado today are accurate to a few parts in a million (10^{12}). In addition, the NBS people in 1970 measured the frequency of the infrared light from the He-Ne laser ($88,376,245 \times 10^6$Hz), which promises to afford a new yardstick also for time. It is a more accurate time-measuring device than has ever existed, and in particular it is easily reproducible anywhere.

Frequency standards (both audio and radio), and many others, also exist in various places in various forms, and all are carefully safeguarded. The standard for the unit of *mass*, the kilogram, is a cylinder of platinum-iridium alloy kept by the International Bureau of Weights and Measures at Paris. This is really the only base unit still defined by an artifact.

To be available for wide use, however, standards must be copied with great care and distributed. From these copies, working standards are then obtained and widely dispersed, as exemplified in a tape certified for length by the NBS in Washington for an individual who desires such assurance. Even so, these "working" standards are not generally used directly for measurements, but are merely used for comparison with working tools or measuring devices.

In our modern world, more measurements are made and are made more accurately than ever before in history. Measurement and a knowledge thereof must be regarded as basic to technology of any kind, thus vital to our whole civilization. New demands have developed rigorous requirements for accuracy, reliability, and sensitivity of measurements. The slower and more tedious measurement methods have been supplanted by swift and frequently complex indirect methods.

1.8 Direct and Indirect Measurements

Direct comparison with a primary or a secondary standard is quite common in length measurement. Analytical chemists use the beam balance to measure (actually, to compare) masses, such measuring being called weighing. A watch is a direct measuring device for time.

More frequently, however, measurements are made through indirect comparison with a standard. Thus a spring balance is used to determine weight by permitting a measurement of the length that a steel spring expands under tension; the usual tire gauge gives an indication of air pressure in much the same way; and a mercury thermometer measures temperature by converting volumetric thermal expansion of the fluid into (essentially) a lineal dimension for viewing. Viscosity of a fluid can be measured by timing the fall of a steel ball through a column of the fluid; the quantity of fluid flowing through a pipe can be found by reading two pressure gauges at different points along the pipe; the compaction of soil can be measured by a nuclear gauge, which counts the rate of reflectance of atomic particles beamed into the soil, and so on. In fact, the whole array of instruments in the modern airplane displays varying measured electrical inputs to indicate fuel quantity, air speed, pressures, temperatures, direction, rates of flow, and so on, many of which are only indirectly measurements of the basic quantities.

Ultimately, however, every measurement must result in a reading or "read-out," since someone at some time must receive the knowledge of a particular comparison to a standard. This reading and its attendant difficulties are the subjects of the ensuing chapters.

Two

Measurement Errors

2.1 Readings

Generally, when readings of any graduated scale are taken, an estimation is made for the final digit—an estimation of the distance between fine scale graduations, such as 6.27 in Fig. 2.1. This could be the end of a 50-ft steel tape graduated in tenths, with the half-tenths also marked. Likely, though, it is a rod reading taken for elevation of the point on which the base of the rod is held.

Note that 6.27 would be the estimate of most people, not 6.26 or 6.28, although these figures almost surely would be estimated by some people in

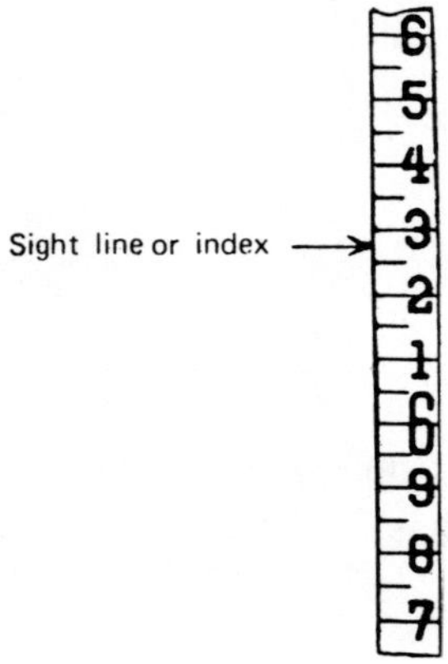

Fig 2.1 Final digit estimated.

a series of measurements of the same quantity. Obviously, if the extra care were warranted, a scale with very fine graduations (say, to thousandths) might be used, and the readings made with a magnifier, probably to ten thousandths of a foot. Such readings, if repeatedly made by the same (or even a different) person, might vary more widely in the last digit (estimated), but it must be remembered that such readings made to ten thousandths instead of to hundredths are more exact than those of Fig. 2.1, and the apparently wider fluctuation in the last place is not nearly so serious.

2.2 Repeated Readings

Assume a series of observed readings for the case above, using the fine graduations and a magnifier, thus reading to ten thousandths of feet (or inches), as in Table 2.1. If the readings had been made to thousandths only, all would have been listed as 6.276; if to hundredths, as 6.28; if to tenths only, as 6.3 units. When a set of such observations shows no variation or very little, one may suspect that the observations or measurements are rough or coarse.

Table 2.1 Readings

1		6.2763
2		6.2757
3		6.2761
4		6.2760
5		6.2761
6		6.2758
7		6.2760
8		6.2764
9		6.2759
10		6.2760
	Mean	6.27603

In this case the best value obtainable from the set is the *arithmetic mean*, commonly called the *average*. This shown as the value 6.27603 or perhaps as 6.2760 units.

It is never possible to obtain absolute correct fourth or fifth decimal-place value in the given instance, simply because the method of measuring was not sufficiently refined. Absolutely correct value does exist, but we could not discover it. What we strive to attain by refined measurements and by techniques of successive measuring is what might be termed a *best available value*.

2.3 Best Value

Because the average in the example is the "best available value," we utilize it as being fairly reliable for the purpose at hand. Though we do not know that it is incorrect, neither do we have any absolute assurance that it is correct.

What is correctness? Since, for the example shown, we cannot discover a true value, we are forced to assume that the mean value (and, in fact, any of the 10 measured values) contains an error. In this context, "error" means difference between measured value and true (or correct) value.

It follows, therefore, that we can never discover a "true error" for any measurement. We shall return to this important concept subsequently.

In the above-mentioned set of measurements, two types of error are present that influence the result:

1. Systematic or constant errors (e.g., temperature).
2. Accidental errors (such as those made in setting the zero end).

2.4 Mistakes

The concept of mistakes is omitted entirely from this discussion; mistakes or blunders are not errors and are never called such. This is not to admit that it is unnecessary, however, to guard constantly against mistakes by frequent checking, since mistakes are caused by inattention or carelessness on the part of the observer.

2.5 Discrepancy

Discrepancy is a term indicating difference between two or more measurements of a quantity. The existence of a discrepancy is frequently a helpful indication of the need for more careful observations. It is not so obvious that more precise measurements are more likely to show discrepancies than cruder ones, but an example will illustrate.

Example. A 9×12 ft rug may be rather casually measured by a housewife, more to determine whether it is a 9 or a 12 ft rug rather than a 10×14 ft rug; no one may notice that it is 9.1×11.8 ft unless the measurement is done more carefully, say, to discover if the rug can run wall to wall in a 9.0×12.0 ft room. Thus more careful measurement would reveal a discrepancy between actual and nominal size. Similarly, a careful machinist may measure a cylinder bore as 3.006, 3.009, 3.004, 3.006, and 3.004 in. on successive measurements to ascertain its roundness; a more casual measurement perhaps would reveal only that the bore is 3.0 in., with no noticeable discrepancy therefrom in many such casual measurements of several diameters.

Thus more precise methods tend to magnify discrepancies or, stated differently, cruder methods tend to hide discrepancies.

SYSTEMATIC OR CUMULATIVE ERRORS

2.6 Systematic Errors

Systematic errors, previously called cumulative, must now be considered. A *systematic error* is one that invariably has the same magnitude and the same sign under the same given conditions. Thus a tailor's cloth tape that has been stretched about 5% by overuse will consistently measure a 40-in. waistline as just over 38 in., flattering the customer but providing a tight fit. Once the condition is known, however, the remedy (or correction) can be applied by adding 1 in. to a 20-in. measurement, $1\frac{1}{2}$ in. to a 30-in. measurement, 2 in. to a 40-in. measurement, and so on.

2.7 Types of Systematic Error

Systematic errors, then, are attributable to known conditions and vary with these conditions. Such errors can be evaluated and applied, with signs reversed, as corrections to measured quantities. They are of three types: natural, instrumental, and personal.

2.7.1 Natural Errors

Natural errors arise from natural phenomena. They are really the effects of certain influences that operate to prevent the observer from seeing or reading directly the quantity being sought. Some instances are the refraction of light rays, the thermal expansion of materials, and the influence of atmospheric pressure or humidity. For example, an instrument that measures distance by timing a radar-frequency radio wave between two points will give an erroneous distance reading if an adjustment (correction) is not made for the effect of barometric pressure and moisure content in the air on the speed of travel of the radio wave.

2.7.2 Instrumental Errors

Instrumental errors are the effects of imperfections in the construction or adjustment of the instruments used in making the measurements. Instances are the lack of concentricity of transit circles, graduation errors in scales, less than perfect optics in a telescope, inertia lag of a needle, worn bearings, and maladjustment of the bubble tube. For example, a spirit level, used to determine the relative elevation of two points, may have the bubble tube axis slightly out of parallel with the sighting axis (line of collimation), giving an erroneous result if the two points are not equally distant from the instrument used for sighting.

2.7.3 Personal Errors

Personal errors depend on the physical limitations and also on the habits of the observer, who may have an auditory lag in noting a time signal, a slight tendency to observe to the right (or to the left) in estimating tenths, or poorly coordinated vision. The amount of such error is usually small, though erratic. For example, in lining up a vernier, the observer may have a tendency to see the line at the left as coincident more frequently than the one at the right, or he may have poor ability in noting time at the beginning and at the end of an interval when using a stopwatch.

2.8 Counteracting Systematic Errors

Although systematic errors are generally cumulative, it is sometimes possible to employ precautionary procedures to prevent their accumulating. For example, if the man on the stopwatch lags in perception by 0.50 at both start and finish of a race, he still gets the correct timing. In this case, use the same man to snap both start and finish. If the spirit-level operator sets his instrument equally distant from points *A* and *B*, he still gets the correct difference of elevation between *A* and *B*, since the plus error cancels the minus error.

Mostly, however, a systematic error does accumulate, as when a shrunken yardstick is laid down successively to set out a tennis court, or an automobile odometer registers 1% high for each mile it travels, or (a common instance) a watch gains or loses a second every hour. In such cases, it is apparent that a correction can be applied if the magnitude and sign of the systematic error are known.

Example. A steel rule 10 ft long is used to measure between two calibration points on the floor of an electronics plant. At the time of the measurement the temperature is 52°F, but the tape is true length (10.0000 ft) only at 68°F. The thermal expansion coefficient for the steel of the tape is 0.00000645 ft/ft. If the reported distance is 127.6120 ft, what is the correct distance?

Correction (C_t) per tape length:

$$0.00000645 \times 10.000 \times (68 - 52) = 0.001032 \text{ ft/tape length}$$

Total correction (for 12.7612 tape lengths):

$$0.001032 \times 12.7612 = 0.0132$$

Corrected length:

$$127.6120 - 0.0132 = 127.5988 \text{ ft} \qquad \textit{Ans.}$$

The correction is subtracted because the tape is effectively shortened by the lower temperature, thus giving a reported measurement that is too large. (The opposite reasoning is employed to apply a correction when measuring off the distance from *A* to set a new point *B*.)

2.9 Detecting Systematic Errors

While measuring, it is important to prevent systematic errors from impairing the accuracy of the final result. The possibility that a particular systematic error exists may be detected by careful analysis of the methods employed and (most effectively) by comparison with independent results that are known to be fairly accurate, especially results attained by a different method. Once systematic errors have been detected, correction is not difficult. The detection of such systematic errors, however, depends on the observer's alertness and knowledge of the natural, instrumental, and personal factors that can influence his procedures.

ACCIDENTAL OR RANDOM ERRORS

2.10 Accidental Errors

Accidental errors or errors of observation are random; they are usually small, and then have a tendency to be mutually compensating. Their presence is indicated in a series of measurements by the appearance of discrepancies. Accidental errors may be either plus or minus; in fact, the probability that the sign is plus is equal to the probability that it is minus. Furthermore, it cannot be determined just what the sign is, since there is no relationship known between the sign and the magnitude of the error on the one hand and any conditions of measurement on the other. They are truly random in occurrence and size.

2.11 Discrepancies Indicativeof Accidental Errors

Only by studying the discrepancies that occur among repeated measurements of the same quantity is it possible to learn anything about the accidental errors inherent in the measurements (except to know that they are bound to exist). For discrepancies (and the consequent residuals) will occur according to a fairly regular pattern when we assume that the instrument used is more refined than the ability of the observer to observe, and that the observations are made with commensurate care and precision.

Example. Suppose an extensometer is used to determine a distance between two gauge marks on a piece of steel. On the main gauge is a reading of 2.14, and on the fine dial are the readings in Table 2.2 (thousandths of an inch). The digit in the column to right of the decimal point is the value that is estimated by the observer (ten thousandths of an inch). Note that the discrepancies from the mean (residuals) vary up to 0.2 division, and be aware that these may be attributable to several kinds of accidental error: setting the points into the gauge holes, imperceptible friction, minor variations of temperature, possible vibration, and so on.

Table 2.2 Gauge Readings

1		6.8
2		7.1
3		6.9
4		6.7
5		6.7
6		6.9
7		6.8
8		7.1
9		7.0
10		6.9
	Mean =	2.1469

2.12 Use of the Mean

Again, where we assume that all the observations are made with the same care and under the same conditions, there is no reason to prefer one observation over any other. Hence all can be said to be of equal weight, and the arithmetic mean (the unweighted mean) is truly regarded as the best obtainable from the observed quantities. The mean is not, of course, the true value, but only the nearest approximation thereto, subject to improvement or change if other observations become available. For the moment it can be adopted as "correct." (This adoption of the arithmetic mean as the best value obtainable from these observed quantities is a sound and fundamental assumption.)

regarded as our probability curve for this experiment. This we shall now explain.

All kinds of random happenings follow the same distribution law or pattern as the pencil drops in this example. If we were to repeat the experiment, it is very probable that our points would fall under the same curve. The area under the curve is finite and, since this entire area represents the probability that a pencil shot will fall somewhere, this area represents certainty and may be taken as 1 or unity. Then, the probability that a pencil drop will fall between any two given limits on the curve is equal to the area under our probability curve between those limits. Also, the number of pencil drops that will probably occur between those limits is the proportional part of the total area of the curve that lies between those limits. This is merely to say that the pencil drops, given no change in the manner of conducting the experiment, will follow the same pattern as before. This ties in with the presentation in Appendix B.

3.5 Normal Distribution Curve

To understand something about distributions, we can examine the histogram of the sum of two dice tossed many times; Fig. B.1 shows it in probability form (Appendix B). The distribution curve is linear and triangular. The plots of the sums of three, four, five, or more dice will become progressively nonlinear (curved) and will, in fact tend to ultimately look like the bell-shaped curve of Fig. 3.6. This curve is the frequency-density

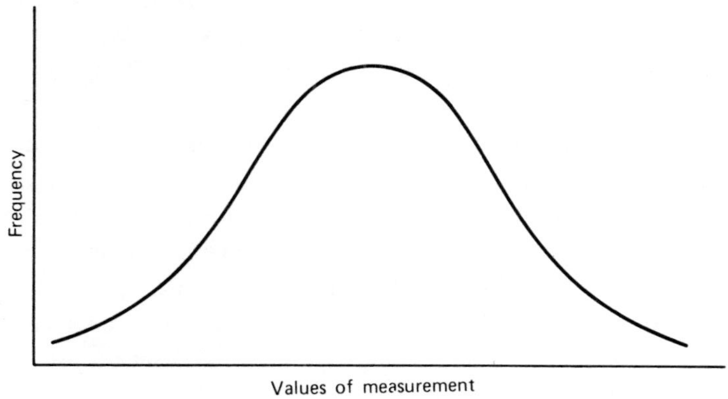

Fig 3.6 Normal distribution curve.

curve for a normal distribution and is known as the *normal distribution curve*, or *normal probability curve*.

Statisticians would insist that the dice throwing results in a Poisson distribution, but for practical purposes the fine distinction between Poisson and normal distribution is best left to more advanced study.

Whether one plots dice sums or pencil drops or weights of 10-year-old boys, the distribution curve tends to be bell shaped, and the relationships of the individual observations to the mean tend to show certain stable characteristics. Whenever one makes multiple measurements of a quantity, the variations from the average have these traits:

1. Small variations from the target (or mean) value occur more frequently than larger ones.
2. Positive and negative variations of the same size are about equal in number, rendering the curve symmetrical about a y-axis.
3. Very large variations seldom occur.

In fact, these three general observations can be noted about virtually any natural phenomenon—for example, the weights of apples gathered from the same tree, the heights of 100 people encountered on the street, the sizes of salmon swimming upstream to spawn, the heights of wheat stalks in a field. The same is observable in manufacturing processes—in the variations in the diameter of ball bearings, in the strength of drawn steel rods, in the bounce of tennis balls, and so on. In engineering measurements likewise it is noticed in connection with such variables as those encountered in the measurements of distances or angles or differences of elevation, of areas by use of a planimeter, of strains in a structural member under tension or compression, of water flow in a pipe or channel, of turbidity or chemical or biological oxygen demand in wastewater. In each case the values tend to follow a normal distribution.

Phenomena like automobile arrivals at an intersection, gaps in traffic streams, and the like, tend to follow a Poisson distribution. So too do other time-independent phenomena like the incidence of telephone calls through an exchange or ship arrivals at a port, the incidence of floods, the location of flaws in materials, and the position of particles of aggregate in surrounding matrices of materials. These are essentially random occurrences and are not particularly the type of engineering measurement occurrences discussed in this book.

It becomes increasingly clear that with large numbers in a natural or controlled grouping the resulting distribution of observed values will be similar to that in any other such grouping. This is true. The frequency-density curve takes on a "normal" appearance, symmetrical and bell shaped. Happily, virtually all engineering measurements one might encounter will have normal distributions. Statistical mathematicians have determined that this "normal" distribution gives rise to a *normal probability curve*, which is a log function of the form $y=ce^{x^2}$, and it is derived specifically as $y=(h/\sqrt{\pi})e^{-h^2x^2}$ in Appendix B.

3.6 Confidence Inspired by Normal Distribution

At this stage it suffices for us to know that for a number of measurements (our *sample*), if our frequency-density curve is shaped like the normal frequency-density curve of the whole *population* (which is the normal probability curve), we can have a measure of confidence. We can begin to believe that our sample group of measurements can be admitted to standard statistical treatment. We can apply the rules of normal probability. We can say of the mean value of our sample that it is the best available value, that our variations from the mean give us a clue to our uncertainty (thus to our certainty), and other helpful things that are characteristic of an infinite group of measurements.

Because we cannot make infinite numbers of measurements each time we need to measure a value, we make a few. Then we examine the distribution of our few results (our sample) to see if it is similar to the (normal) distribution curve of the entire population. We can do this simply enough at this juncture by plotting the histogram and curve of our sample to study its shape (Appendix B). More advanced statistical analysis does afford mathematical tests for normalcy of data. Once we accept that our sample data are normally distributed, we can presume that our sample group is part of the whole family (population) of such measurements and may be treated as such.

This is not a cause for misgivings. Instead, the ascribing of qualities of exactness and the assigning of degrees of certainty to our smaller group of measurements, simply by noting the set's resemblance to an "infinitely" large set, is a matter of economics. It is also safe to do. The very resemblance of the small set's distribution of error (shape of the curve) is a matter of confidence attained through similarity.

3.7 Skewed Frequency-Density Curve

It is apparent that the pencil drop curve of Fig. 3.5 is leaning a little toward the right (tailed or skewed a little toward the left), thus making the normality (or normalcy) of the distribution there plotted somewhat doubtful. This stems, it seems reasonable to conclude, from the fairly limited number of values plotted. Given a near-infinity of values, the curve should plot up as entirely symmetrical.

If we have a badly skewed curve, however, we ought to examine our measurements and probably remake them. A hidden systematic error may be having a telling effect on our work. Serious skew would indicate that plus variances are much larger in size than minus variances, or vice versa, and the distribution is not normal.

As an example, golf balls sold in a pro shop under the best name have been tested for bounce (e.g., over a barrier), and those rejected are sold under a less prestigious name. A test of a batch of the best balls may well show a distribution skewed to the right, since the tail to the left will not exist (Fig. 3.7).

Incidentally, all the ensuing work is predicated on the normal distribution of results. This is brought out further in Chapter 9.

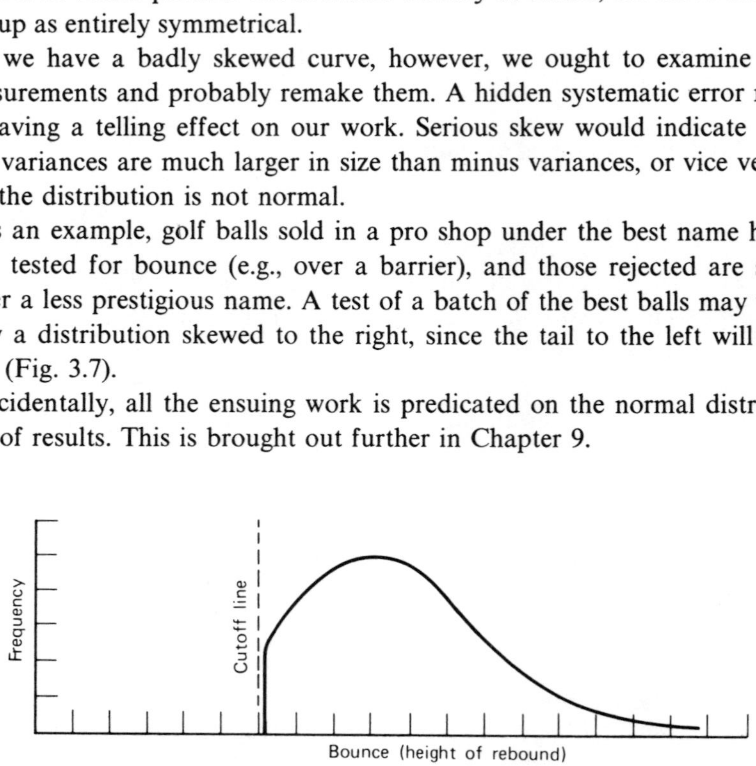

Fig 3.7 Distribution of top name golf balls.

3.8 Least Squares Principle

If a set of measurements M_1, M_2, $M_3, \ldots,$ is made, all with equal care, and using the same methods and instruments for each, the precision constant h is the same for all measurements, and the distribution of their random errors $(x_1, x_2, x_3, \ldots)$ will be given by the same probability curve. The probabilities of the occurrence of these errors then, are

$$p_1 = \frac{h}{\sqrt{\pi}} e^{-h^2x_1^2} dx_1$$

$$p_2 = \frac{h}{\sqrt{\pi}} e^{-h^2x_2^2} dx_2$$

The derivation of this is in Appendix B.

3.9 Errors and Residuals

It is important at this point to distinguish between two concepts that are used interchangeably, namely, error and residual. The absolute magnitude of a quantity can never be determined by measurement because of the presence of accidental (random) errors. Obviously, then, the error of a measurement can never be determined, since an error is the difference between the measured value and the true value. In practice we use the best value of a series of measurements, namely, the arithmetic mean, in place of the (unknown) true value. The difference between the arithmetic mean and any particular measurement is called the *residual* for the measurement. The arithmetic mean is used, therefore, as a representation of the true value.

The arithmetic mean is truly representative of the series of measurements (and, therefore, equal to the true value) when the distribution of random errors is completely uniform. However this occurs only when the number of measurements is infinite. Obviously, this is an impossible limitation: we always necessarily work with fewer than this infinite number of measurements.

Hereafter, the error is denoted by x and the residual (variation) by v. It is, of course, evident that since we cannot know the true size of the (accidental) errors (x), we shall most frequently be using residuals or variations from the mean (v) in our formulations. Since the separate measurements are independent events, the probability that the whole set of errors ($x_1, x_2, x_3, \ldots$) will be made is the product of their separate probabilities. That is,

$$P = p_1 \cdot p_2 \cdot p_3, \ldots$$

$$= \left[\frac{h}{\sqrt{\pi}} e^{-h^2(x_1^2 + x_2^2 + x_3^2 + \ldots)} \right] (dx_1, dx_2, dx_3, \ldots)$$

Because in any set of measurements a set of small errors is more probable than a set of large errors, the set that has the greatest probability gives us the best or most probable value of the quantity measured. Since the precision index h is constant and the differentials $dx_1, dx_2, dx_3,, \ldots,$ are arbitrary quantities, it is evident from the equation above that the probability P is greatest when the exponent of e is the least, that is, when $x_1^2 + x_2^2 + x_3^2, \ldots,$ is a minimum —when Σx^2 is a minimum.

Thus the principle of least squares can be expressed as:

The best or most probable value obtainable from a set of measurements of equal precision is that value for which the sum of the squares of the errors is a minimum.

The requirement that the sum of the squares be a minimum makes it evident that the arithmetic mean is the best value obtainable from any set of direct measurements that are equally trustworthy.

This means simply that if the arithmetic mean ("average") of a set of several measurements is used as a value to represent the set, it is a better value than any other value because the sum of the squares of the difference of each value from the mean will be the least possible.

Example. Given a set of mileage distances between two points as read from the odometers of 10 cars, calculate the Σv^2 if $v = X_i - \bar{X}$ then, letting $\bar{X}'$ be any value except the average or mean, again calculate Σv^2 and compare.

	X_i	v	v^2	v'*	v'^2
1.	104.8	−0.1	0.01	−0.2	0.04
2.	106.2	+1.3	1.69	+1.2	1.44
3.	103.7	−1.2	1.44	−1.3	1.69
4.	104.5	−0.4	0.16	−0.5	0.25
5.	104.2	−0.7	0.49	−0.8	0.64
6.	104.9	0.0	0.00	−0.1	0.01
7.	104.1	−0.8	0.64	−0.9	0.81
8.	105.2	+0.3	0.09	+0.2	0.04
9.	106.2	+1.3	1.69	+1.2	1.44
10.	105.2	+0.3	0.09	+0.2	0.04
	$\bar{X} = 104.9$		6.30		6.40

*Let $\bar{X}'$ arbitrarily be 105.0 and $v' = X_i - \bar{X}'$.

It is seen that $\Sigma v'^2 = 6.40$ (larger than 6.30) when a value (than the mean is used to calculate the sum of the squares of If a value lower than the mean were used (say, 104.8), the Σ be larger than 6.30 (actually, by coincidence, also 6.40 in tha The following discussion shows the justification for using variations from the mean.

3.10 Equating Residuals with Accidental Errors

If for a finite number of measurements we assume that

$$\bar{X} = \bar{X}_0 - \frac{\Sigma x}{n}$$

(arith. mean) (true value) (mean error)

it can be seen that since the errors (x) are as likely to be positive as negative, the quantity Σx is not large, and $\Sigma x/n$ is still smaller. Hence the larger the number of measurements, the closer the arithmetic mean ($\bar{X}$) approaches the true value ($\bar{X}_0$), which is also the population mean.

The pattern of occurrence of (accidental) errors is assumed to follow that of the residuals from the mean. Thus if we assume that (in Fig. 3.8)

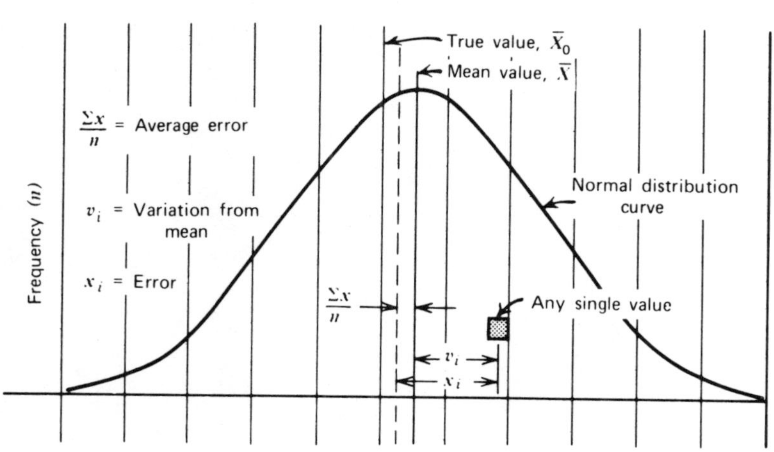

Fig 3.8 Relationship of error and residual.

$$\underset{\text{(any residual)}}{v_1} = \underset{\text{(corresponding error)}}{x_1} - \underset{\text{(mean error)}}{\frac{\Sigma x}{n}}$$

it is apparent that when the number of measurements is large, residuals are almost equal to the errors. Hence although we can never determine the true magnitude of a measured quantity, we can determine it as closely as we please by taking enough measurements. In practice, therefore, there is always some uncertainty in the determination of the true magnitude of a quantity. This uncertainty is an estimate of the precision of the measurement, and the precision in turn is an estimate of the accuracy of the measurement (since we have eliminated systematic errors from consideration by applying corrections for them).

3.11 Precision an Indicator of Accuracy

Because the arithmetic mean and its associated residuals give a poor indication of true value when the number of measurements is small, we postulate that the precision of any large set of measurements is an estimate of their accuracy. However we have seen that as the number of measurements is increased, the value of the residuals approaches that of the errors. Therefore, if a sufficiently large number of measurements is employed, the residuals may be replaced by the errors, and vice versa. Thus it is seen that if we narrow the value of the arithmetic mean to a very small range by using residuals, we must conclude that we are also narrowing it down to a very small range with respect to errors, thus approaching very close to a true (accurate) value.

Four

Probability Theory of Errors

4.1 Mean Square Error

The question now arises how to estimate the uncertainty (hence the precision and, therefore, the accuracy) of a measurement. We can use as a measure of the precision any one of several devices, but here we shall speak of the mean square error (the square root of the mean of the squares of the errors). The *mean square error* is defined (see Section 3.9) as being equal to

$$\sqrt{\frac{x_1^2+x_2^2+x_3^2+\ldots+x_n^2}{n}}=\sqrt{\frac{\Sigma x^2}{n}}$$

This is the mean square error of any single observation of the set.

4.2 Standard Deviation

Since we cannot know the values of errors $x_1, x_2, x_3, \ldots$, we use values that we can ascertain, the residuals $v_1, v_2, v_3, \ldots$. First, however, let us insist on a refinement, so as to be in accord with statistical authorities.

We have seen previously that

$$\underset{\text{(any residual)}}{v_1} = \underset{\text{(corresponding error)}}{x_1} - \underset{\text{(mean error)}}{\frac{\Sigma x}{n}}$$

By working this relationship into the equation of the *normal probability curve*, although it is beyond the scope of this treatment, it is possible to establish that the *mean square error* (dealing with errors) and what we call the *standard deviation* (dealing with residuals) are not quite equal. Whereas the mean square error of any single measurement in a set if $\sqrt{\Sigma x^2/n}$, the standard deviation for any single measurement is defined as

$$\sigma_s = \sqrt{\frac{\Sigma v^2}{n-1}} = \sqrt{\frac{v_1^2 + v_2^2 + v_3^2 + \ldots}{n-1}}$$

Despite the slight difference, however, the term *mean square error* is frequently used for σ_s instead of the more properly termed *mean square variation*, or *mean square deviation*, or simply *standard deviation* (of any single measurement in a set).

To call σ_s the standard deviation of a "single measurement" is a bit misleading, because the term "standard deviation" is characteristic of the distribution of the sample or the set, not of an individual measurement. It cannot be found for a single measurement that is not part of a group. As a clarification, sometimes it is said that σ_s describes the next measurement of the set that might be made (or, in fact, any one of them that has already been made). The expression describes an individual measurement made under the sample circumstances, attributing to the individual value the quality that derives from the group (sample) of measurements. This is much as each child in a family shares in the family's reputation.

It may also be noted that sometimes the value σ_s is shown as $\sqrt{\Sigma v^2/n}$ instead of $\sqrt{\Sigma v^2/(n-1)}$. The difference between the two becomes less important as the value of n increases. There is no real harm, once the number of measurements becomes 10, 15, or more, in using either of these formulas. This, too, will explain why it is possible to speak of *mean square error* and *standard deviation* as synonymous and equivalent.

4.3 Standard Deviation of the Population

Besides knowing from the normalcy of our sample data that the sample mean is very close to the mean of the population, we have another assurance. Statisticians tell us that the standard deviation of the sample (σ_s) is related to the standard deviation of the whole population (S.D.) as follows:

$$\sigma_s \text{ (sample)} = \frac{\text{S.D. (population)}}{\sqrt{n}}$$

where n is the number of values in the sample.

As an example, the notion could be used in tenuous fashion for a clue to the spread in values that might be expected from a new nuclear density device before the manufacturer has had opportunity for exhaustive study. With one instrument some 20 measurements could be made to calculate the σ_s of the sample. From this the standard deviation of the whole population of such measurements (S.D.) could be estimated. So, if σ_s were calculated to be ± 8.5,

$$\text{S.D. of population} = \sigma_s \sqrt{n} = \pm 8.5\sqrt{20} = \pm 38.0$$

Subsequently, through myriad tests (a "near infinity"), the manufacturer would arrive at a surer knowledge of the S.D. of the population, which should be close to ± 38.0.

Lest concern over this arise, it must be said that no great use is made of the S.D. of the population. It becomes helpful in describing the capability of an instrument or a procedure, however, in some situations. This relationship is not of great import and should not be allowed to confuse the present situation. The S.D. of the whole population can never be known without an infinite number of measurements, generally speaking, but this relationship is a means of finding it approximately.

4.4 Number of Values in a Sample or Set

Sets of measurements thus far have been comprised of 10 values, 20 values, or 845 pencil drops. In statistics, is 10 a small number, or 20? Or would a sample of 10 or 20 be large enough? Basically if the 10 or 20 measurement values (or any larger number) form a good distribution pattern and appear to be a good sample of the entire normal population of such measurements, we can accept them as a sample. It can be ascertained that we would not be very likely to get greatly different values no matter how many more times we try. So a sample of 10 or 20 or so would surely then be large enough.

In this context, numbers like 100, 200, and so on, are surely large; numbers like 10, 15, or 20 seem small but with caution can be used. As a rule of thumb, it may be well to regard 20 as a sort of dividing line. Experience is the best teacher in this respect.

4.5 Standard Error

We frequently wish to know something about the uncertainty of the arithmetic mean. It may be reasoned that the uncertainty of the arithmetic mean of a series of measurements is much less than that of any single measurement. In fact, it has been determined (though not here) that the *mean square error* of the arithmetic mean of a set is properly the mean square error of any single value divided by the square root of the number of measurements. We do not speak of this as the "standard deviation" but call it the *standard error* (of the mean). Thus the *standard error* of the arithmetic mean equals the standard deviation of any single measurement divided by the square root of the number of measurements.

$$\sigma_m = \frac{\sigma_s}{\sqrt{n}} = \sqrt{\frac{\Sigma v^2}{n(n-1)}} \approx \sqrt{\frac{\Sigma v^2}{n(n)}} = \frac{\sqrt{\Sigma v^2}}{n}$$

We shall denote standard error by σ_m and standard deviation by σ_s. The σ_s value may be variously called *sigma error*, the *mean square error*, the *root mean square error*, or the *standard deviation*. It must be remembered as the square root of the mean of the squares of the deviations from the mean (or variations from the mean). The standard error σ_m is sometimes referred to as the *sigma error of the mean*, confusingly enough. Care must be taken to avoid mixing the two, standard deviation and standard error.

Withal, for the present purpose we shall use these relationships and terminology unless otherwise indicated or implicit from the context:
Standard deviation of a set or sample:

$$\sigma_s = \sqrt{\frac{\Sigma v^2}{n-1}}$$

Standard error of a set or sample:

$$\sigma_m = \sqrt{\frac{\Sigma v^2}{n(n-1)}} = \frac{\sigma_s}{\sqrt{n}}$$

4.6 Meaning of the Standard Error

Assume that one set of measurements has been made. The mean ($\bar{X}$) of the several values is calculated, the value best descriptive of the sample. The standard deviation (σ_s) of the sample is also found, which is truly descriptive of the spread and the distribution of the values in the sample. This is true because the sample is presumed to have been carefully chosen as a representative sample and is expected to have a normal distribution characteristic running through it. Thus σ_s is descriptive of where the individual values fall within the sample, and it is descriptive of the "normality" of their spread.

But to describe the mean value ($\bar{X}$) we use the standard error (σ_m). This relies on the value of n also: larger values of n tend to lessen this error of the mean. By carrying the reasoning to its limit, we see that as n approaches infinity, the σ_m approaches zero and the mean of the sample must approach the mean of the entire population of such measurements.

So the standard error can be seen to be indicative of the nearness of the sample mean to the population mean, thus to true value. This welcome achievement, the ability to quantitize our nearness to correct value, is the basis for our confident use of less than an infinite number of measurements—in fact, of a relatively few. This, then, is the key to "statistical measurement."

4.7 Use of the Standard Error

Another notion follows from the foregoing. Assume that several samples (or sets of measurements) are taken, each sample properly representative of the population under study. Then the mean of these several sets forms a distribution whose standard deviation is called the standard error (of the sets). It can be reasoned that the average of the mean of two sets will be better than the mean of only one sample; in like manner, the mean of the

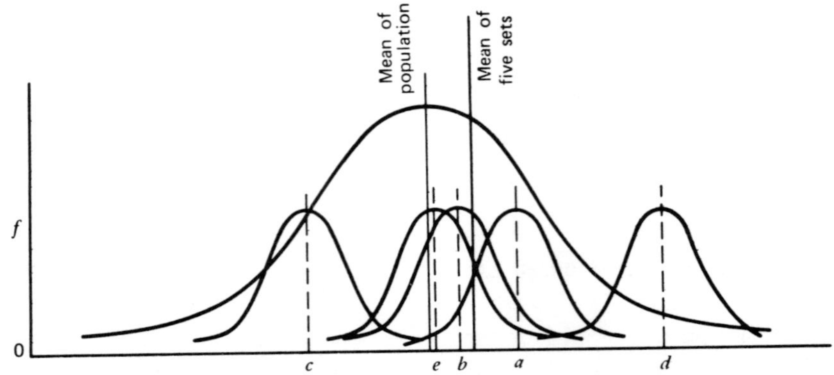

Fig 4.1 Relationship of samples to population: *a–e* are the means of sets of individual measurements.

means of several sets can confidently be expected to approach more nearly the mean of the entire population.

The reasoning is obvious if one considers taking measurements in such sets until the whole population (the "infinite" number of measurements) has been utilized. One must assume that the means of the several sets of measurements are normally distributed within the population, an assumption that can hardly be denied.

Then the mean values of each of these several sets will constitute a normally distributed set of mean values, which will in turn have a standard deviation and a standard error. There may be two, three, or many sets. The multiple sets will each fall within the population, and so, too, will their mean. Moreover, the mean of all the sets will fall closer to the population mean than any individual mean (it can be confidently expected). And the standard error will indicate how near the mean of all the means will be to the population mean. Figure 4.1 illustrates the principle.

Standard error is further used in the later chapters; Section 8.13 gives the evaluation of the mean.

4.8 Frequency Distribution Table

The mean, standard deviation, and standard error of the mean can best be calculated in tabular form, as in this and later examples illustrate. In sets of many measurements a tally sheet will prove convenient, as at the end of Appendix F.

Example. The 10 distance measurements that follow are made, for which standard deviation, standard error, and best value are desired.

X (m)	v	v^2
68.161	−0.0005	0.00000025
68.162	+0.0005	25
68.161	−0.0005	25
68.163	+0.0015	225
68.160	−0.0015	225
68.162	+0.0005	25
68.164	+0.0025	625
68.161	−0.0005	25
68.160	−0.0015	225
68.161	−0.0005	25
681.615	0.0000	0.00001450
ΣX	Σv	Σv^2

Mean:

$$\bar{X}=\frac{\Sigma X}{n}=\frac{681.615}{10}=68.1615 \text{ ft}$$

Standard deviation:

$$\sigma_s=\sqrt{\frac{\Sigma v^2}{n-1}}=\sqrt{\frac{0.00001450}{9}}=\pm 0.00127 \text{ m}$$

$$\text{or } \pm 0.001 \text{ m}$$

Standard error of the mean:

$$\sigma_m=\frac{\sigma_s}{\sqrt{n}}=\pm\frac{0.00127}{\sqrt{10}}=\pm 0.0004 \text{ m}$$

Best value:

$$68.1615 \pm 0.0004 \text{ m}$$

4.9 Short Method for Standard Deviation

For convenience in calculation, the mean may be assumed at the start and corrected later. This method is well adapted to desk calculator or electronic computer, but it is especially helpful in manual computation. The following examples use a simplified computation with an assumed mean ($\bar{X}'$), which is adjusted to get the correct mean ($\bar{X}$). Any variation from the assumed mean is labeled v'.

$$\bar{X} = \bar{X}' + \frac{\Sigma v'}{n}$$

Then the standard deviation is found thus:

$$\sigma_s = \sqrt{\frac{n}{n-1}}\sqrt{\frac{\Sigma v'^2}{n} - \left(\frac{\Sigma v'}{n}\right)^2}$$

or approximately (but safely) by this formula:

$$\sigma_s \approx \sqrt{\frac{\Sigma v'^2}{n-1} - \left(\frac{\Sigma v'}{n}\right)^2}$$

Example 1. By this assumed mean method, the work of the example of Section 4.8 is rendered somewhat easier and simpler, especially for hand calculation. Assuming that $68.160(\bar{X}')$ can serve as an initial estimate of the mean, variations (v') will work out conveniently and the mean can be later adjusted:

$X(m)$	v'	v'^2
68.161	+0.001	0.000001
68.162	+0.002	4
68.161	+0.001	1
68.163	+0.003	9
68.160	+0.000	0
68.162	+0.002	4
68.164	+0.004	16
68.161	+0.001	1
68.160	+0.000	0
68.161	+0.001	1
($n = 10$)	+0.015	0.000037
$\bar{X}' = 68.160$ (assumed)	$\Sigma v'$	$\Sigma v'^2$

Mean:

$$\bar{X}=\bar{X}'+\frac{\Sigma v'}{n}=68.160+\frac{0.015}{10}=68.160+0.0015=68.1615\text{ m}$$

Standard deviation:

$$\sigma_s=\sqrt{\frac{n}{n-1}}\sqrt{\frac{\Sigma v'^2}{n}-\left(\frac{\Sigma v'}{n}\right)^2}=\sqrt{\frac{10}{9}}\sqrt{\frac{0.000037}{10}-\left(\frac{0.015}{10}\right)^2}$$

$$=\pm(1.0541)(0.001204)=\pm 0.00127 \quad\text{or}\quad \pm 0.001\text{ m}$$

By approximate formula, we have $\sigma_s=\pm 0.00136$ or ± 0.001 m.

A further simplification of the computation can be accomplished, especially with large samples, by grouping the data in ascending order and using a frequency factor, much according to the following distribution tabulation (see form of Appendix F). Again we use $\bar{X}'=68.160$:

X (m)	f	v'	fv'	v'^2	$f(v'^2)$
68.160	2	0.000	0.000	0.000000	0.000000
68.161	4	1	4	1	4
68.162	2	2	4	4	8
68.163	1	3	3	9	9
68.164	1	4	4	16	16
	$(n=)$ 10		0.015		0.000037

Mean:

$$\bar{X}=\bar{X}'+\frac{\Sigma fv'}{n}=68.160+\frac{0.015}{10}=68.1615\text{ m}$$

Standard deviation:

$$\sigma_s=\sqrt{\frac{n}{n-1}}\cdot\sqrt{\frac{\Sigma f(v'^2)}{n}-\left(\frac{\Sigma fv'}{n}\right)^2}$$

$$=\pm 0.00127\text{ m} \quad\text{or}\quad \pm 0.001\text{ m}$$

Example 2. From the ages of a group of children here given, find the average age, the standard deviation, and the standard error of the mean age.

15	12	11	8	13
13	12	8	14	15
9	10	12	11	11

By the usual method:

Age(s)	f	fx	v	v^2	fv^2
8	2	16	−3.6	12.96	25.92
9	1	9	−2.6	6.76	6.76
10	1	10	−1.6	2.56	2.56
11	3	33	−0.6	0.36	1.08
12	3	36	+0.4	0.16	0.48
13	2	26	+1.4	1.96	3.92
14	1	14	+2.4	5.76	5.76
15	2	30	+3.4	11.56	23.12
11.60	15	174			69.60

$$\therefore \text{mean} = \frac{\Sigma fx}{\Sigma f} = \frac{174}{15} = 11.60$$

(from which find v in each case)

$$\sigma_s = \sqrt{\frac{\Sigma fv^2}{n-1}} = \sqrt{\frac{69.60}{14}} = \pm 2.23$$

$$\sigma_m = \pm \frac{2.23}{\sqrt{15}} = \pm 0.58$$

By the assumed-mean method: assume that the mean is some middle convenient value, say, 10:

Age	f	v'	fv'	$(v')^2$	$f(v')^2$
8	2	−2	−4	4	8
9	1	−1	−1	1	1
10	1	0	0	0	0
11	3	+1	+3	1	3
12	3	+2	+6	4	12
13	2	+3	+6	9	18
14	1	+4	+4	16	16
15	2	+5	+10	25	50
	15		+24		108

Mean:

$$\bar{x} = \bar{x}' + \frac{fv'}{f} = 10.00 + \frac{24}{15} = 10.00 + 1.60 = 11.60$$

Standard deviation:

$$\sigma_s = \sqrt{\frac{n}{n-1}}\sqrt{\frac{\Sigma fv'^2}{n} - \left(\frac{\Sigma fv'}{n}\right)^2}$$

$$= \sqrt{\frac{15}{14}}\sqrt{\frac{108}{15} - (1.60)^2} = \pm 2.23$$

Standard error:

$$\sigma_m = \frac{\pm 2.23}{\sqrt{15}} = \pm 0.58$$

Example 3. An angle was measured by observers *a*, *b*, and *c*. The values reported (in each case the mean of several measurements) were 63°14′10.5″, 63°14′11.0″, and 63°14′12.0″, respectively. Given these measurements (seconds of arc) with a theodolite, find the standard deviation of the group, and the mean.

Observer	X	f	fX	v	fv	fv^2
a	10.5″	2	21.0	−0.72	2(−0.72)	1.04
b	11.0″	4	44.0	−0.22	4(−0.22)	0.20
c	12.0″	3	36.0	+0.78	3(0.78)	1.81
Σ		9	101.0			3.05

Mean:

$$\bar{X} = \frac{\Sigma fx}{n} = \frac{101.0}{9} = 11.22''$$

Standard deviation:

$$\sigma_s = \sqrt{\frac{\Sigma fv^2}{n-1}} = \sqrt{\frac{3.05}{8}} = \pm 0.62''$$

Standard error:

$$= \frac{0.62}{\sqrt{9}} = \pm 0.21''$$

We may note the mean value and the "accuracy label" of the angle can be written: angle is 63°14′11.22″ ±0.21″.

Example 4. A set of 439 observations was made on a distant rod target. The results shown are again tabulated in a manner adapted to easy plotting and to easy computation of the mean. Again, a frequency table is used, and the mean is calculated in the first three columns. Then, to obtain the third and fourth columns, the decimal and front zeros are omitted. To obtain the fifth column, rather than multiply 6.571 by 1, 6.572 by 8, and so on, we shorten the work and reduce the chance of error by mentally subtracting 6.500 from each X value (again omitting the decimals for the time being). The 6.500 is restored just beneath the table. Also, to calculate the residuals, the mean ($\bar{X}$) is rounded off to 6.578 in this instance.

Value, X (ft)	Number of occurrences, f	Product, fX (omit 6.500)	Residual (or variation), v (thousandths)	fv	fv^2
6.571	1	71	−7	−7	49
6.572	8	576	−6	−48	288
6.573	18	1,314	−5	−90	450
6.574	27	1,998	−4	−108	432
6.575	36	2,700	−3	−108	324
6.576	43	3,268	−2	−86	172
6.577	53	4,081	−1	−53	53
6.578	55	4,290	0	0	0
6.579	53	4,187	+1	+53	53
6.580	46	3,680	+2	+92	184
6.581	36	2,916	+3	+108	324
6.582	26	2,132	+4	+104	416
6.583	15	1,245	+5	+75	375
6.584	13	1,092	+6	+78	468
6.585	7	595	+7	+49	343
6.586	2	172	+8	+16	128
6.5782	439	34,317			4,059
(mean)	n	ΣfX			Σv^2

Mean:

$$\frac{\Sigma fX}{n} = \frac{34{,}317}{439} = 78.17 \qquad \text{or} \qquad 6.500 + 0.07817 = 6.5782 \text{ ft}$$

Standard deviation:

$$\sigma_s = \sqrt{\frac{\Sigma v^2}{n-1}} = \sqrt{\frac{4059}{438}} = \sqrt{9.267} = \pm 3.04$$

and since we are here dealing with thousandths of a foot,

$$\sigma_s = \pm 0.00304 \text{ ft} \quad \text{or} \quad \pm 0.0030 \text{ ft} \quad \text{or} \quad \pm 0.003 \text{ ft}$$

This may be called the standard deviation from the mean for any single measurement. Also the *standard error of the mean* is

$$\sigma_m = \frac{\sigma_s}{\sqrt{n}} = \pm \frac{0.00304}{\sqrt{439}} = \pm 0.00015 \quad \text{or} \quad \pm 0.0002 \text{ ft}$$

4.10 Use of Standard Errors

Standard error is a useful device for comparing different sets of measurements of the same quantity, and is simply illustrated as follows. Suppose that the set of 439 measurements (Section 4.9) is to be compared with two other sets of measurements

Set	n	Mean	Standard error, σ_m
A	439	6.5782	± 0.00015
B	167	6.5784	± 0.00033
C	702	6.5778	± 0.00010

It can be seen that set C has the highest precision and set B the least. If a *best value* is to be found by using all three sets, the middle set should be given the least weight and the last set the greatest weight. Weights are discussed in Chapter 8.

4.11 Plotting the Histogram and the Frequency-Density Curve

The plot of the values from Example 4 of Section 4.9 is shown in Fig. 4.2 with $\pm \sigma_s$ ordinates drawn. The histogram (rectangles) has been plotted,

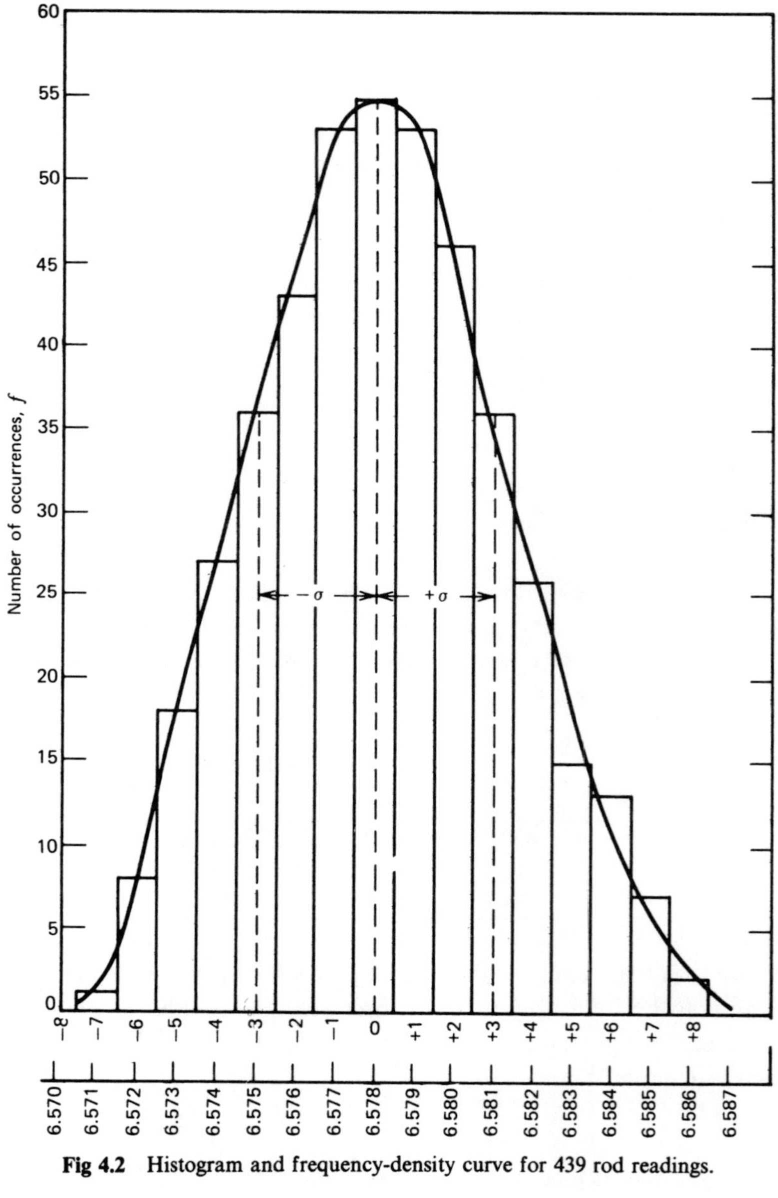

Fig 4.2 Histogram and frequency-density curve for 439 rod readings.

and the tops of the rectangles are connected by a smooth bell-shaped curve. It is almost the typical probability curve. The probability curve can be described as a continuous curve that applies continuous variables where the difference between one value and the next can be indefinitely small. (Appendix B gives a slightly different plot for the same values, where the *normal probability density curve* or *normal distribution curve* is plotted on the histogram in Fig. B.4.)

Connecting the midpoints of the tops of the histogram bars gives a jagged frequency-density curve for any sample unless the number of values is very large. As the number approaches "infinity," the curve tends to be smooth, symmetrical, and bell shaped. Sometimes a rather smooth curve for fewer values is faired in approximately through the proper points to convey the distribution picture—perhaps more truly than may the jagged curve, since the phenomenon itself would have a smooth distribution by its very nature.

Notice, too, that a curve more nearly approaching the bell shape of the probability curve would have resulted if more than 439 values had been used. Mathematically, the curve stretches to infinity in each direction, becoming asymptotic. However, only the small portion shown is important to us.

Plotted on the curve of Fig. 4.2 are the $\pm\sigma_s$ values, corresponding to actual rod readings of 6.5752 and 6.5812 ft. The area beneath the curve between these two ordinates represents the number of observations that fell between the two values. From a quick estimate of the number of values that fall between 6.5752 and 6.5812 (from the frequency distribution table of Example 4 of Section 4.9), it is seen that about 68% of the value do fall between $+\sigma_s$ and $-\sigma_s$. Further refined analysis, especially of sets of measurements containing many more values, would bear this out. Reference to Table 5.1 will indicate that there is a 68.3% certainty that any single value, selected at random from among these measurements of a set, will fall within the $\pm\sigma_s$ range (i.e., will be a value that is within ± 0.000304 ft of the mean, 6.5782 ft). Chapter 5 discusses the probability curve and its meaning as a measure of reliability of measurements.

In the histogram of the pencil drops (Fig. 3.4) either narrower boxes or wider boxes might have given a different shape to the distribution curve —and to the histogram. Had only 10 been used instead of 20, the pattern probably would have been sufficiently apparent and just as useful. Usually, about 10 to 14 boxes will do.

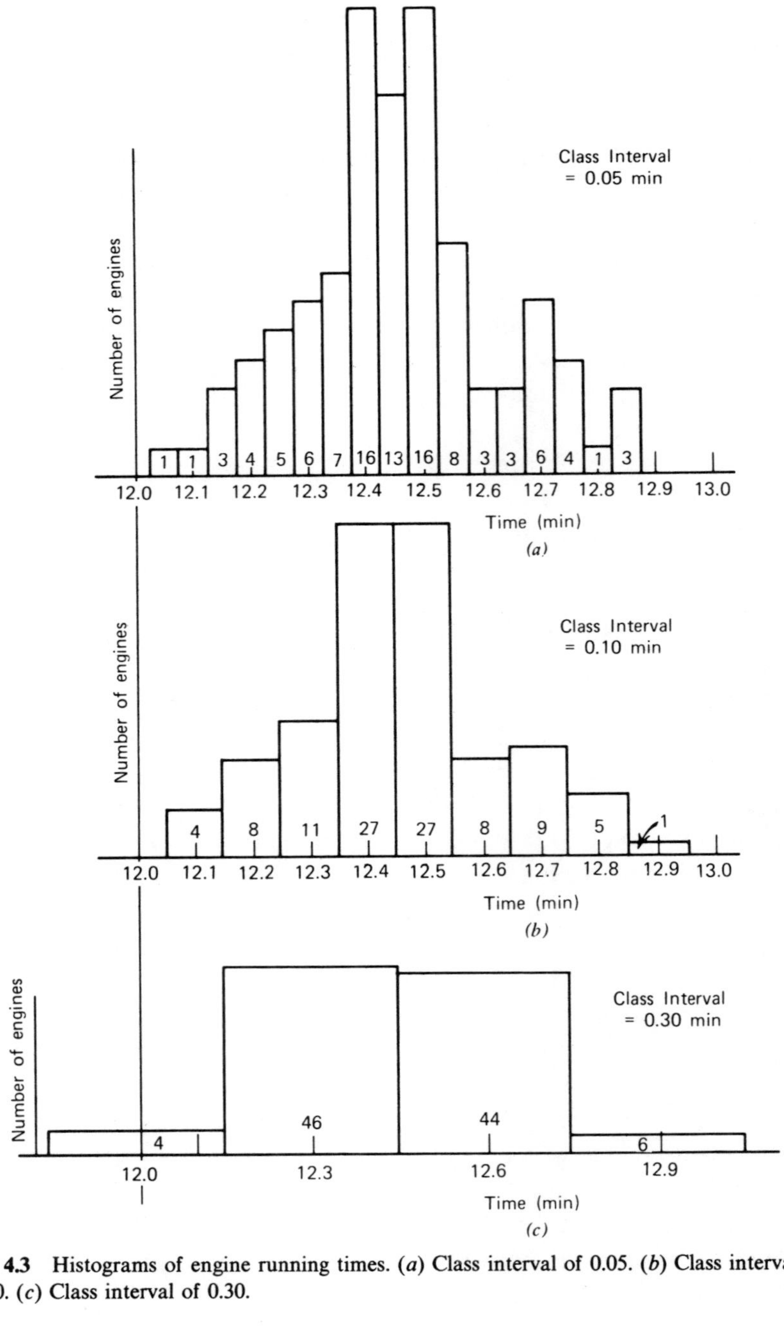

Fig 4.3 Histograms of engine running times. (*a*) Class interval of 0.05. (*b*) Class interval of 0.10. (*c*) Class interval of 0.30.

An illustration follows, using the engine running times, on a test bench, of 100 selected engines; this shows that using different class sizes or *class intervals* can create different impressions with the histogram. Choosing the class interval too narrow (causing too many histogram bars) can spread the histogram badly and rob it of significance. Choosing too wide a class interval (thus too few classes) can obliterate the histogram pattern (see Fig. 4.3)

Example. On the test bench automobile engines are run under unvarying conditions to discover how long each one runs with a given, measured amount of fuel. One hundred successive engines were timed as follows:

Automobile engine runs (min)

12.54	12.68	12.43	12.64	12.14
12.46	12.40	12.49	12.37	12.42
12.37	12.69	12.38	12.66	12.41
12.10	12.42	12.40	12.48	12.38
12.40	12.72	12.35	12.44	12.18
12.39	12.60	12.16	12.50	12.21
12.45	12.39	12.39	12.19	12.46
12.74	12.51	12.23	12.47	12.40
12.62	12.71	12.45	12.24	12.44
12.21	12.45	12.47	12.46	12.49
12.52	12.29	12.06	12.74	12.84
12.36	12.50	12.29	12.53	12.37
12.58	12.14	12.42	12.48	12.79
12.23	12.50	12.39	12.68	12.53
12.43	12.42	12.76	12.51	12.46
12.49	12.52	12.56	12.51	12.57
12.87	12.33	12.48	12.33	12.53
12.31	12.42	12.85	12.52	12.76
12.68	12.28	12.64	12.29	12.30
12.53	12.48	12.27	12.56	12.26

Grouped with a class interval of 0.05 minutes, we have the following:

Class	Class interval	f
12.00	11.076–12.025	0
.05	12.026–12.075	1
.10	12.076–12.125	1
.15	12.126–12.175	3
.20	12.176–12.225	4
.25	12.226–12.275	5
.30	12.276–12.325	6
.35	12.326–12.375	7
.40	12.376–12.425	16
.45	12.426–12.475	13
.50	12.476–12.525	16
.55	12.526–12.575	8
.60	12.576–12.625	3
.65	12.626–12.675	3
.70	12.676–12.725	6
.75	12.726–12.775	4
.80	12.776–12.825	1
.85	12.826–12.875	3
.90	12.876–12.925	0
		$\Sigma = 100$

Grouped with a class interval of 0.10 min, we have

Class	Class interval	f
12.0	11.96–12.05	0
.1	12.06–12.15	4
.2	12.16–12.25	8
.3	12.26–12.35	11
.4	12.36–12.45	27
.5	12.46–12.55	27
.6	12.56–12.65	8
.7	12.66–12.75	9
.8	12.76–12.85	5
.9	12.86–12.95	1
13.0	12.96–13.05	0
		$\Sigma = 100$

Grouped with a class interval of 0.30 min, we have

Class	Class interval	f
12.0	11.851–12.150	4
.3	12.151–12.450	46
.6	12.451–12.750	44
.9	12.751–13.050	6
13.2	13.051–13.350	0
		$\Sigma = 100$

4.12 Normal Probability Distribution Curve

The bell-shaped curve is the normal probability *density* curve, but its use as a tool is rather limited. Frequently, a simple variant of this curve can be used, namely, the normal probability *distribution* curve. This is given in Appendix F, together with simplified methods of plotting data for a ready visual test of the data on special arithmetic probability paper.

Five

Measures of Reliability

5.1 Definitions

A few definitions are now in order because some of the terms that are frequently used in statistical work have definite meanings. The number of times a variable occurs in a set of observations is called the *frequency* (*f*) of the occurrence of the value. A value that represents a series of observations may be the *average*, a term that does not invariably designate the arithmetic mean but simply the *central tendency*, where the values tend to cluster.

5.2 Mean

The best known and most useful "average" is the *arithmetic mean*, usually referred to as the *mean*; it is calculated by adding all observations and dividing the sum by the total number of observations. In Example 4 of Section 4.9, the mean is 6.5782 (or 6.578 if rounded off).

5.3 Median

Another "average" often used is the *median*. It is the middle observation, or the arithmetic mean of the two middle observations. The median represents the set of observations in the sense that the number of observations greater than this value is the same as the number that are smaller. In a good, symmetrical set of values, the median will equal the mean. For instance, in a set of 439 observations (Example 4, Section 4.9) the median will be the 220th value, easily found by counting off 220 values in the f column—provided the values have all been tallied as shown in columnar form. It is 6.578.

5.4 Mode

Another "average" sometimes used is the *mode*, that is, the most frequently occurring value. In Example 4 of Section 4.9 the mode is 6.578, since it occurs 55 times, more than any other. If a graph of the frequency

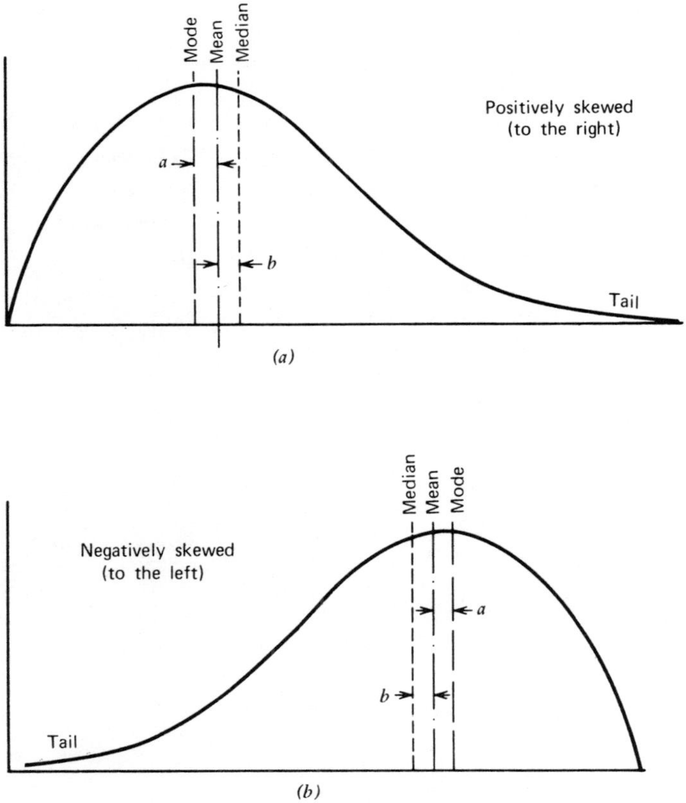

Fig 5.1 Skewed distribution. (*a*) Positively skewed (to right). (*b*) Negatively skewed (to left).

distribution is drawn, the mode is the high point or hump of the curve (see Fig. 4.2). If the curve is symmetrical, it is easily seen that the mode, the mean, and the median are the same value. But in a curve that is skewed (or tailing off) to the right or left, the mode, mean, and median will be somewhat separated.

A test that involves mean, mode, and median can be made for skewness in a distribution curve. As a rule of thumb, the distribution is regarded as moderately skewed when the difference between mean and mode (a) is approximately three times the difference between mean and median (b), as Fig. 5.1 shows. It may also be observed from the diagram that:

1. The *mean* passes through the centroid of the area beneath the curve.
2. The *median* divides the area into equal halves.
3. The *mode* is fixed by the highest point on the curve.

5.5 Scatter

The way that the different values lie about this average is called the dispersion or *scatter*, which is, of course, the chief indicator of precision of the set of measurements. Scatter is used to study certainty and uncertainty. Scatter of the values in a set of observations is an indication of their reliability. Wide dispersal bespeaks less reliable data than observations that lie closely distributed about the mean. The shape of the bell curve (the probability curve) gives an indication of scatter: a flat curve would mean greater scatter than a tall narrow curve.

5.6 Range

The simplest measure of scatter is *range*, the distance between the largest and the smallest value. The range of the values in Example 4 of Section 4.9 is 0.017 ft, the difference between 6.571 and 6.586. However range is less indicative of reliability than are certain other measures.

5.7 Mean Deviation

More reliable for measuring scatter is *mean deviation* (average error), obtained by adding the individual variations for each value (from any average), and dividing by the number of values. Thus we can have a mean deviation (or mean variation) from the mean, from the median, or from the mode.

In Example 4 of Section 4.9 the sum of the fv column in the tabulation (Σfv) divided by 439 (n) gives:
Mean deviation:

$$\frac{\Sigma fv}{n} = \frac{1075}{439} = 2.45 \text{ thousandths}$$

Interpreted properly, this shows that ± 0.00245 ft is the average error or the mean variation or deviation from the mean (average) of the rod readings.

5.8 Variance and Standard Deviation

The mean deviation (average error) has a weakness because it tells nothing about the manner in which the values are dispersed. Indeed, one set of measurements may have errors all of medium size, and another set of measurements may have errors of variable size: several very large, a few medium, and several very small. Yet both sets of measurements may have the same average error. Comparison of the mean deviations of the two sets would give a false indication: in the mean deviations there would be no indication of the distribution (scatter) of the values in each set. There would be no penalty for large errors and no reward for small ones. Fortunately, a better and more universally used measure of scattering is the *standard deviation*. This is, once again, the square root of the mean of the squares of the deviations (variations) of the observations from their arithmetic mean. It is usually symbolized by small sigma (σ_s), and is also frequently called the *mean square error*.

Another measure of scatter is *variance*, defined as

$$\sigma_s^2 = \frac{\Sigma fv^2}{n-1}$$

which is seen to be the square of the standard deviation.

The *coefficient of variance* is also sometimes used, defined as a ratio of the variance to the mean of a given sample or set ($= \sigma_s^2/\overline{X}$). It has some meaning when comparing standard deviations, as seen in this example.
Voltage measurement A:

$$\sigma_s = \pm 0.5 \text{ V}$$

Voltage measurement B:

$$\sigma_s = \pm 20 \text{ V}$$

Unless a comparison were made using the coefficients of variance; one might conclude (erroneously) that the A measurement is much better. If $\bar{X}_A = 6.00$ V and $\bar{X}_B = 14{,}400$ V, then:

$$\frac{\sigma_A^2}{\bar{X}_A} = \frac{0.5^2}{6.00} = 0.042 \qquad \text{or} \qquad 4.2\% \text{ variance}$$

$$\frac{\sigma_B^2}{\bar{X}_B} = \frac{20^2}{14{,}400} = 0.028 \qquad \text{or} \qquad 2.8\% \text{ variance}$$

It is then seen that the B measurement is proportionately better.

The *coefficient of variation* is used sometimes in this situation for comparison. It is defined slightly differently, as $V_x = \sigma/\bar{X}$ and could serve the same purpose as the coefficient of variance.

A word of caution must be interjected here lest one be tempted to compare blindly all measurements in this manner. It must at once be interjected that this apparent equivalence can mislead one who may be tempted to compare blindly all measurements in this manner. It would not be valid for barometer readings, map coordinates, distance readings on a steel tape, angle measurements, or even readings at various scale positions on a voltmeter. In such instances, the value of the quantity read is extraneous. To compare by the coefficients of variation a direction of 5° with a direction of 275° being read with a theodolite is obviously wrong, since the mechanics of reading the theodolite are identical for each case. The magnitude of the direction (angle) does not affect the instrument's accuracy or its capability for precision. To use coefficients of variance on temperature readings of +5°F and +85°F would obviously be wrong as well. One must be careful not to misapply coefficient of variance; it has but limited application.

5.9 The Sigma Error

The error σ_s is used synonymously with "uncertainty," and it has been found to be of such size that 68.3% of the errors in the series are less than it and 31.7% are greater. This means that there is a 68.3% certainty that the error of any single measurement will fall between $+\sigma_s$ and $-\sigma_s$. Thus it may be said that a measurement in a series whose deviation from the mean value, either positively or negatively, is greater than σ_s will occur only about 1 in 3 times in the series.

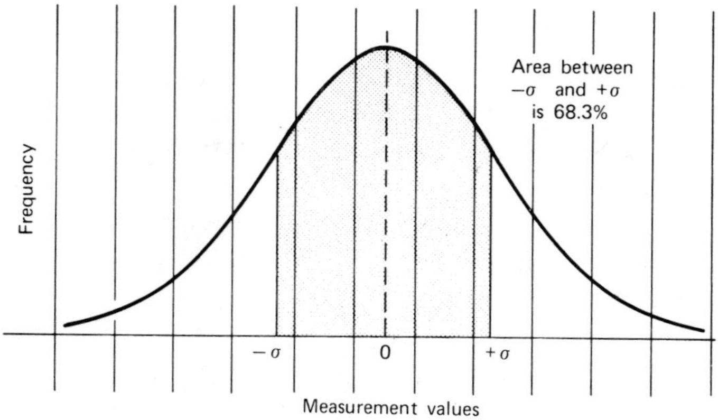

Fig 5.2 Incidence of errors within one sigma of the mean.

It can be seen, therefore, that on the normal probability curve the area contained between the $-\sigma_s$ and the $+\sigma_s$ values of error will amount to 68.3% of the total area. It means that 68.3% of the errors will be plotted between $-\sigma_s$ and $+\sigma_s$, as shown in Fig. 5.2.

5.10 The Two-Sigma Error

Further consideration reveals that the area between $-2\sigma_s$ and $+2\sigma_s$ will contain about 95.5% of the total area. This means that a measurement whose positive or negative deviation from the mean value is greater than $2\sigma_s$ will occur only about 1 in 20 times; or $\frac{19}{20}$ of the time the error of a single measurement in the series will fall within $\pm 2\sigma_s$, as Fig. 5.3 indicates.

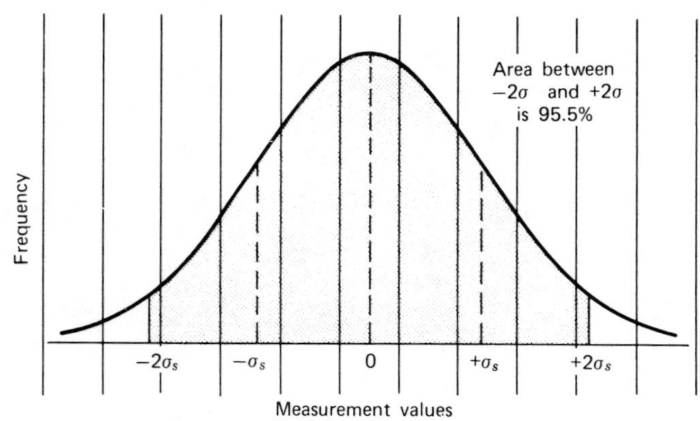

Fig 5.3 Incidence of errors within two sigma of the mean.

5.11 The Three-Sigma Error

Finally, the area between $-3\sigma_s$ and $+3\sigma_s$ on the normal probability curve will contain about 99.7% of the total area. This means that a measurement whose deviation from the mean value, either positively or negatively, is greater than $3\sigma_s$ will occur only about 1 in 370 times; or 369/370 of the time the error of a single measurement in the series will fall with $\pm 3\sigma_s$, as Fig. 5.4 shows.

5.12 Probable Error

"Probable error" was once greatly emphasized, but it is being used less and less, mainly because it is defined essentially as a 50% certainty, or 50–50 chance. On the curve of Fig. 5.5 the *probable error* is shown by ordinates, between which exactly half of the errors occur.

The value of the probable error (E_p) is 0.6745 σ_s. It can be said of any single measurement that its deviation from the mean has an equally good

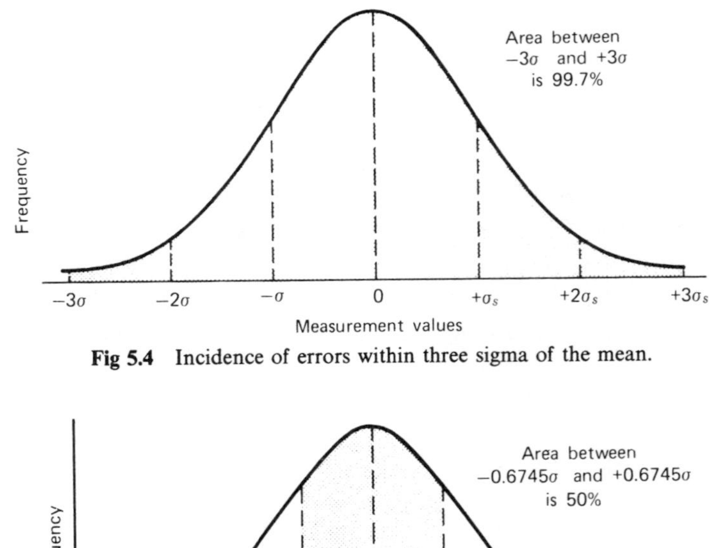

Fig 5.4 Incidence of errors within three sigma of the mean.

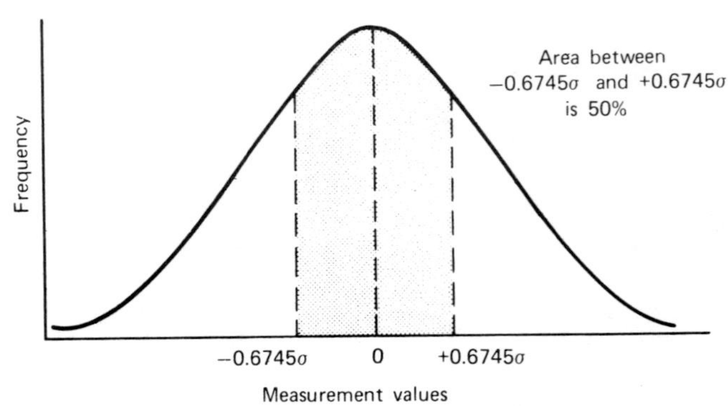

Fig 5.5 The probable error or 50% error.

(50–50) chance of being greater or less than the probable error value. However fewer and fewer people care to know their chances of the 50–50 occurrence, preferring to know their certainty of being within a known value of the mean on 68.3, 95.5, or 99.7% of their observations.

5.13 Meaning of Standard Deviation and Standard Error

The standard deviation (σ_s) tells us that any individual measurement (maybe a next measurement) in our sample or set has a 68.3% probability of lying within $\pm\sigma_s$ of our sample mean, a 95.5% probability of being within $\pm 2\sigma_s$, and so on. This follows once we have established that there are no systematic errors remaining and that the sample is normally distributed, for we know the mean has a tendency to be centrally located in the sample population.

The standard error (σ_m) tells us that the set itself has that particular precision as a set, for a small σ_m means a tightly grouped sample, and so on. But it also means that another similar sample of the same population would have a 68.3% probability of lying within $\pm\sigma_m$ of the present sample (and vice versa). So, with many other similar sets or samples of the same population, each would have a 68.3% chance of being within $\pm\sigma_m$ of the other sets in the population, of any and all other such sets in the population. The logic of the situation compels us then to believe that the σ_m of *any* such sample (including our own) has a 68.3% probability of falling within $\pm\sigma_m$ of the mean of the means of all the sets, and also of the mean of the whole population of these measurements. This gives certitude (within the percent confidence limits) of being within some known distance of the *true* value. And this certitude stems from just a single set or sample of measurements.

Our standard error computation thus leads directly and very logically to a certitude for our measurement work. And we can say of the result of our measuring even a single set, even a small sample, that we can be 68.3% sure, 95.5% sure, or 99.7% sure that our mean value lies within $\pm\sigma_m$, $\pm 2\sigma_m$, or $\pm 3\sigma_m$ of the true value of a quantity. This is virtually the culmination of our entire study of error theory. This makes statistical measurement (a valid phrase) a very cogent tool for the engineer.

Figure 5.6 shows the areas contained in the various sections of the normal distribution curve (or probability curve). The abscissa would be σ_s for one sample or set of measurements, but σ_m for a very large sample or for a group of several samples or sets of measurements.

In any event, the significant point is that our level of confidence that the

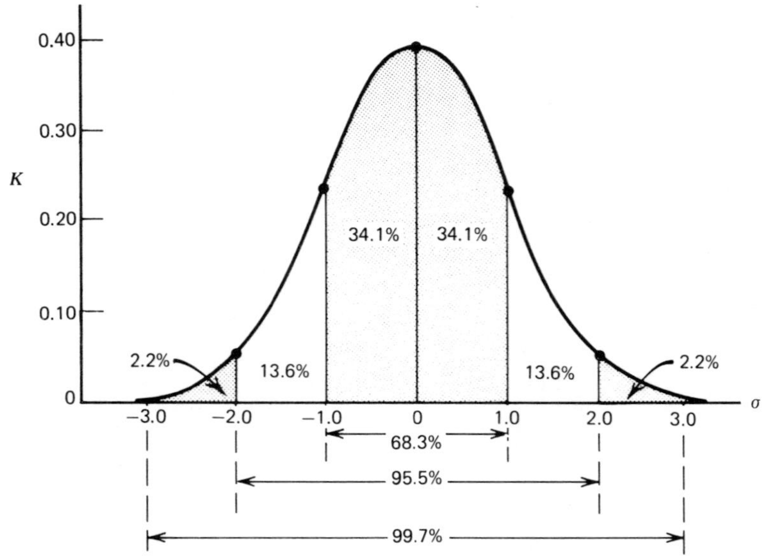

Fig 5.6 Characteristics of the normal distribution curve.

sample mean lies within $\pm\sigma_m$ of the population mean is 68.3% (only 1 chance in 3 of lying beyond); within $\pm 2\sigma_m$, 95.5% (only 1 chance in 20 of lying beyond); and so on. Table 5.1, the summary table, shows other key values. It is cautioned that these values are strictly valid only if the sample size is 20 or more; they should be used with some care if the sample is smaller. Nevertheless, this ability to establish some sort of confidence from making multiple measurements is one very great benefit of statistical

Table 5.1 Size of Error in a Single Measurement of a Set: A Summary

Name of error	Symbol	Value	Certainty (%)	Probability of larger error
Probable	E_p	$0.6745\ \sigma_s$	50	1 in 2
Standard deviation	σ_s	$1.0\ \sigma_s$	68.3	1 in 3
90% error	E_{90}	$1.6449\ \sigma_s$	90	1 in 10
Two-sigma or 95.5%error	$2\sigma_s$	$2\sigma_s$ or $3E_p$	95.5	1 in 20
Three-sigma or 99.7% error	$3\sigma_s$	$3\sigma_s$	99.7	1 in 370
Maximum*	E_{max}	$3.29\ \sigma_s$	99.9+	1 in 1000

*Some authorities regard the 95.5% error as the "maximum error." Neither view is absolutely correct, since the theoretical maximum error is $\pm\infty$, which does not occur in practice. It is, then, a good practical decision to use the 95.5% or the 99.9^{+}% error as the "practical" maximum that is tolerable.

analysis. We speak of the 68.3% *confidence interval*, the 90% *confidence interval*, and so on.

Table 5.2 gives handy factors for converting from one to another error size when the occasion arises.

Table 5.3 lists the relationship between various sigma errors in the

Table 5.2 Linear Error Conversion Factors

	To					
From	50.0% E_p	68.3%, σ_s	90.0% E_{90}	95.5%, $2\sigma_s$	99.7%, $3\sigma_s$	99.9$^+$% E_{max}
E_p	1.000	1.483	2.439	2.965	4.449	4.875
σ_s	0.674	1.000	1.645	2.000	3.000	3.288
E_{90}	0.410	0.608	1.000	1.216	1.824	2.000
$2\sigma_s$	0.337	0.500	0.822	1.000	1.500	1.644
3σ	0.225	0.333	0.548	0.667	1.000	1.096
E_{max}	0.205	0.304	0.500	0.608	0.912	1.000

Table 5.3 Area Beneath the Normal Probability Curve

σ_s	(% area)	σ_s	(% area)	σ_s	(% area)	σ_s	(% area)
0.0	50.0	1.0	84.1	2.0	97.7	3.0	99.86
.1	54.0	.1	86.4	.1	98.2	.1	99.90
.2	57.9	.2	88.5	.2	98.6	.2	99.93
.3	61.8	.3	90.3	.3	98.9	.3	99.95
.4	65.5	.4	91.9	.4	99.2	.4	99.97
.5	69.2	.5	93.3	.5	99.4	.5	99.98
.6	72.6	.6	94.5	.6	99.5	.6	99.98
.7	75.8	.7	95.5	.7	99.6	.7	99.99
.8	78.8	.8	96.4	.8	99.7	.8	99.99
.9	81.6	.9	97.1	.9	99.8	3.9	100.00

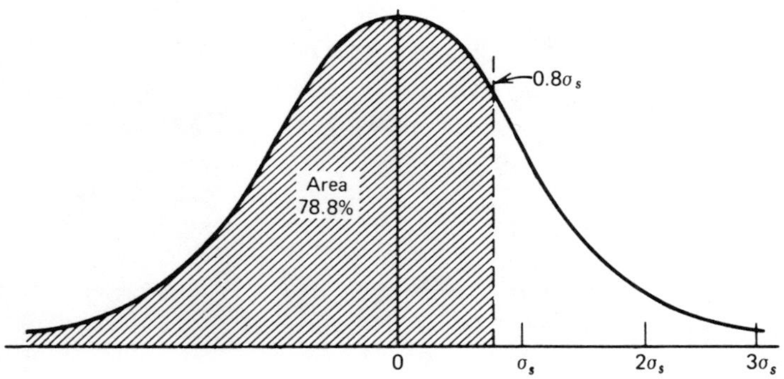

sample and the percentage of values *below* the particular sigma value in a normal distribution. It lends itself to the interpreting of test results, adherence to specifications, and similar procedures. For instance, in Fig. 4.2 it is obvious that 84.1% of all values lie below 6.581; one can figure that 90.3% of all values lie below 6.582, and so on. Similarly, one can see that since 99% of the values lie below 6.588, only about 1% of them exceed this value. It is a handy enough table to find virtually any needed relationships.

Example 1. Here are the results of an angle measurement, using an instrument which reads to 1 second of arc (estimated to tenth of second). Calculate the best value, the mean deviation, the standard deviation, and the standard error.

Value,	X	v	v^2
19° 27′	36.2″	0.42	0.1764
	34.1	2.52	6.3504
	39.7	3.08	9.4864
	40.1	3.48	12.1104
	36.2	0.42	0.1764
	34.1	2.52	6.3504
	35.2	1.42	2.0164
	35.7	0.92	0.8464
	34.9	1.72	2.9584
	37.1	0.48	0.2304
	38.0	1.38	1.9044
	37.2	0.58	0.3364
	37.8	1.18	1.3924
	36.1	0.52	0.2704
	35.9	0.72	0.5184
	36.1	0.52	0.2704
	36.8	0.18	0.0324
	37.9	1.28	1.6384
	34.0	2.62	6.8644
	39.3	2.68	7.1824
mean	36.62	28.64	61.1120
		Σv	Σv^2

Standard deviation

$$\sigma_s = \sqrt{\frac{\Sigma v^2}{n-1}} = \sqrt{\frac{61.1120}{19}} = \sqrt{3.2164} = \pm 01.79''$$

Mean deviation:

$$\bar{v} = \frac{\Sigma v}{n} = \frac{28.64}{20} = \pm 01.43''$$

Best value:

$$\bar{X} = 19° \ 27' \ 36.62''$$

$$2\sigma_s = \pm 03.59''$$

$$3\sigma_s = \pm 05.38''$$

Standard error:

$$\sigma_m = \frac{\sigma_s}{\sqrt{n}} = \pm \frac{01.79}{\sqrt{20}} = \pm 00.40''$$

Note once again the meaning of "sigma error," which is either the standard deviation or the standard error, depending on the context. This time let us apply the standard error (σ_m) to the results of the Example 1. We may say of the value of the angle (presuming, of course, that systematic errors have all been eliminated) that:

1. The most likely value is 19°27′36.62″, which is the arithmetic mean.
2. There is a 68.3% certainty that the true value lies between 19°27′36.22″ and 19°27′37.02″ (i.e., $\pm\sigma_m$).
3. There is a 95.5% certainty that the true value lies between 19°27′35.82″ and 19°27′37.42″ (i.e., $\pm 2\sigma_m$).
4. There is a 99.7% certainty that the true value lies between 19°27′35.42″ and 19°27′37.82″ (i.e., $\pm 3\sigma_m$).

It might also be noted that the E_p or "probable error" (based on 50% certainty) for the sample mean is only $0.6745\sigma_m$ ($= \pm 00.27''$). We can say then that the *most probable value* of the mean angle (only a 50–50 surety) lies between 19°27′36.35″ and 19°27′36.89″ in this case. The fine expressions "most probable error" and "most probable value"—or simply "probable error" and "probable value"—have been usurped to refer only to the 50% values. This is correct, but unfortunate. Thus whenever we wish to designate one or the other of the "sigma errors" or "sigma values," we must be careful not to use the word "probable."

Example 2. Steel rods from a particular batch are tested in tension for yield strength; there are 18 rods in all. The mean strength of the sample is found to be 26,490 psi with $\sigma_s = \pm 1450$ psi. Does this batch satisfy the specification that 95% of the rods shall be at least 25,000 psi?

Note that if one uses the 90% coefficient from Table 5.1, there will be 5% cut off the curve to the left, the failures; the 5% cut off at the right are stronger than required, thus are acceptable.

$$1.645\ \sigma_s = 1.645\ (1450) = \pm 2390 \text{ psi}$$

$$26{,}490 - 2390 = 24{,}100 \text{ psi (below which 5\% of the rods would fall)}$$

So, the batch does not live up to specification, since only about 85% will pass (one can judge from the curve).

Using Table 5.3 to find an exact percentage above 25,000 psi, we have

$$\frac{26{,}490 - 25{,}000}{1450} = \frac{1490}{1450} = 1.028\ \sigma_s$$

Interpolating between the 1.0 and the 1.1 values:

$$0.841 + (0.864 - 0.841)\frac{0.028}{0.10} = 0.847 \qquad \text{or} \qquad 84.5\% \text{ passing}$$

If the 18 rods, however, were but one isolated sample in a vast population, we might validly turn to the standard error (standard deviation of the mean) of this sample and make some predictions about the entire population. The standard error is:

$$\sigma_m = \frac{\sigma_s}{\sqrt{n}} = \frac{1450}{18} = \pm 342 \text{ psi}$$

Then from Table 5.3 we find that the coefficient is 1.645, or 95% will pass the test of $26{,}490 - 1.645\ (342) = 26{,}490 - 562 = 25{,}928$ psi. Thus it can be expected that as a population, some 95% *will* pass the requirement for 25,000 psi. This exemplifies the use of a small sample to describe the larger population.

Example 3. If a probable error (50%) is known to be ± 1.927 mph in a speed reading of 100 mph, find the standard deviation to be expected in a test of speedometers off the assembly line.

From Table 5.2 we find the conversion factor (1.483).

$$\sigma_s = \pm 1.927(1.483) = \pm 2.858 \text{ mph}$$

5.14 Probability Applications

Since in most instances distributions in the general run of engineering problems tend to be normal, the probabilities shown heretofore can be expected. Life expectancy of highways, electrical goods, machines, and so on, then can be fairly well predicted from test performances. Some examples will indicate how practical and powerful a tool—under careful handling—we have now uncovered.

Example 1. In a test of 30 light bulbs sampled from a batch of 1000, the average life tested out to be 1367 hours. About 5% of the sample failed before 1153 hours, and 5% were still burning after 1590 hours. A bar graph (histogram) was kept in the test room from which a distribution curve was drawn; a normal distribution curve was fitted also to the values, from which some predictions were attempted by the manufacturer (Fig. 5.7). Note that the $-3\sigma_s$ mark (99.7%) falls at 975 hours, establishing that only 0.15% (2 of 1000) will fail by then. A quick eye estimate will establish that no more than three or four should fail by 1000 hours. A 1000-hour guarantee seems safe to make.

It is easy to apply Table 5.3 to the bell curve. Note that the table shows 97.7% for $+2.0\ \sigma_s$, which means that 97.7% will burn out by 1590 hours and 2.3% will last beyond. Mentally reversing the picture of the table gives us the story for the $-2.0\ \sigma_s$ line of Fig. 5.7. Now we see that 2.3% will fail before 1100 hours and 97.7% will continue to burn past then. Thus Table 5.3 can serve a double purpose.

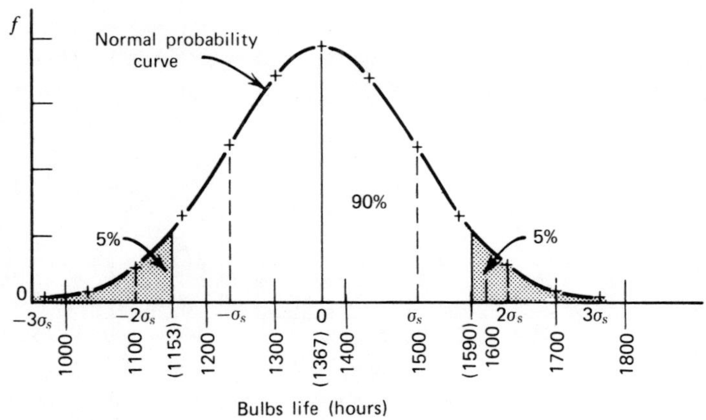

Fig 5.7 Life- span of 1000 lamp bulbs in a batch.

Plotting the lamp bulb data on arithmetic probability paper is rewarding. (See Appendix F.) The mean life is plotted at the 50% mark on the paper. Then the $\pm\sigma_s$ value must be found, and the clue is that 5% failed before 1153 hours and 5% were still burning at 1590 hours, a spread of 437 hours. Or we can find that the 90% spread is

$$\frac{(1367-1153)+(1590-1367)}{2}=218.5 \text{ hours}$$

the average spread each way from the mean. Now by using Table 5.2 to get the conversion factor, we can discover the value of σ_s:

$$\sigma_s = 0.608(218.5) = \pm 133 \text{ hours}$$

This enables us to plot the values for $+\sigma_s$ and $-\sigma_s$ on the paper and draw the straight line, our frequency distribution curve plotted on arithmetic probability paper (Fig. 5.8). We quickly ascertain that 1% (10 lamps) will burn out by 1070 hours, 5% (50 lamps) will burn out by 1153 hours, and so on. The plot on arithmetic probability paper is indeed much simpler and

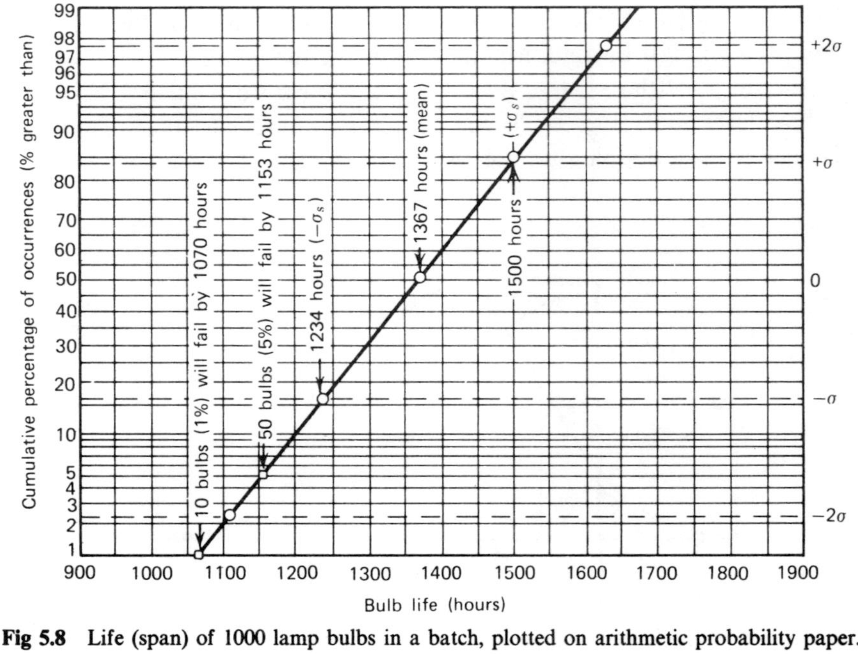

Fig 5.8 Life (span) of 1000 lamp bulbs in a batch, plotted on arithmetic probability paper.

easier and provides basically the same information as does the frequency-density plot (the bell-shaped curve).

Although all the preceding analysis of light bulb life was based on standard deviation (σ_s), it may well be considered that the 30 units were selected in this instance as one of many ongoing sample sets of a continuing line of production. It becomes logical and legitimate then to base our predictions on the standard deviation of the sample mean (which is σ_m or standard error). Drawing a simple line, then, on arithmetic probability paper with the σ_m value as our new basis (Fig. 5.9) we find new results:

$$\sigma_m = \frac{\sigma_s}{\sqrt{n}} = \frac{133}{\sqrt{30}} = 24 \text{ hours}$$

average life (mean) = 1367 hours
99% will burn past 1312 hours (since only 1% will fail by then)
95% will burn past 1328 hours (since only 5% will fail by then)
90% will burn past 1337 hours (since only 10% will fail by then)

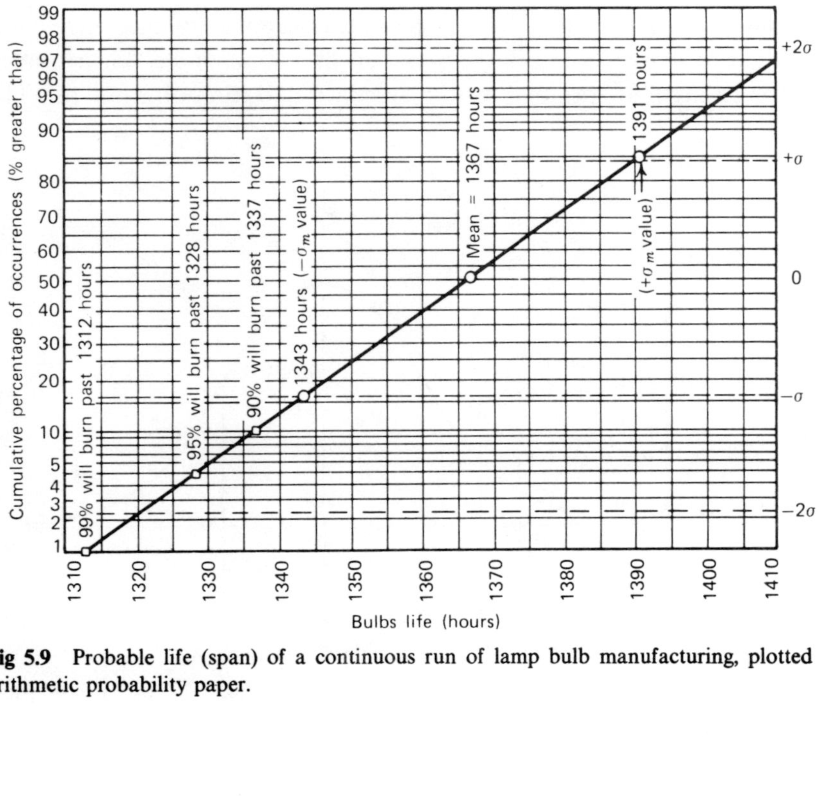

Fig 5.9 Probable life (span) of a continuous run of lamp bulb manufacturing, plotted on arithmetic probability paper.

These are valid predictions predicated on this test and on the normalcy of manufacturing conditions.

If we turn to the obverse situation, we may seek to discover, for example, what strength of concrete pavement will be needed to meet a given specification imposed by state road authorities. Some interesting reasoning must emerge. If the state inspector needs 3500 psi concrete, could he permit a few test cylinders to show only 3450 psi? How many cores at less than 3500 psi can be tolerated? Is it perhaps reasonable to demand that 95% be 3500 psi or better and that none shall be less than 3400 psi?

The road constructor could surely add extra cement to get all his concrete better than 3500 psi, but his bid price might be entirely too high. Yet if he cuts his margin too thin, he may be required to break up and replace a considerable expanse of rejected pavement. So, the answer must be to use available sampling information to predict what a particular combination of cement, water, aggregate, mixing plant, and placement method will produce. Figure 5.10 is a representation of a large sample of tests, with some conclusions that follow.

Example 2. Information from a particular supplier of pavement concrete indicates that for a large number of cylinders he has supplied concrete strength as specified, with a standard deviation of ±325 psi. The

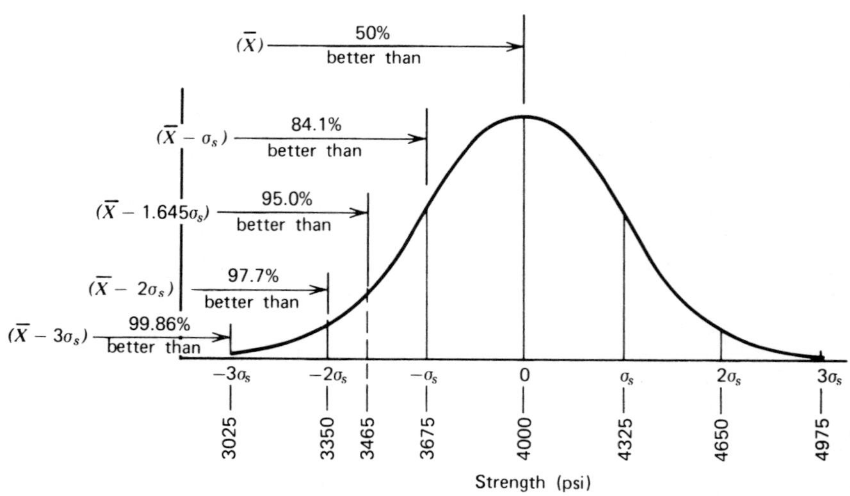

Fig 5.10 Normal distribution for concrete cylinder strengths: σs, ±325 psi; $\bar{X}$, 4000 psi.

engineer for the CBA Construction Company draws the normal distribution curve (normal probability curve) centered about a mean of 4000 psi. He needs to determine how to meet a strength specification of 3500 psi. Plotted on the curve are the σ_s, $1.645\sigma_s$, and $2\sigma_s$ values, along with their respective "percentages better than" (Fig. 5.10).

From the bell curve the engineer sees it is possible to anticipate the several indicated percentages attaining the several strengths as shown. But unless some specific percentage exemption is contained in the specification, he must shift the entire curve to the right. It is obvious that if 99.86% must meet the 3500 psi specification, the curve would need a shift of about 500 psi, to a mean value ($\bar{X}$) of 4500 psi. The "better than" indications on the bell curve stem from Table 5.3.

It is seen that if 5% of the test cylinders were permitted to fall below 3500 psi, one might suggest that the constructor set as his "target" a mean strength of 4035 psi if he wants to deliver acceptable concrete. If it were possible through tighter control in mixing to cut the σ_s from 325 psi to perhaps 200 psi, a saving could result by ordering concrete of a slightly lower strength. (It is quite unlikely that this could be done, however, principally because it is difficult to control the aggregate closely.)

By plotting the design curve (mean = 4000 psi) on arithmetic probability paper, a simpler curve results (Fig. 5.11), and the problem is easily manipulated. Various parallel straight curves can be drawn until one of them satisfies the specifications. For instance, one curve is drawn to intersect the 3500 psi ordinate at 10% (giving a new mean strength of 3910 psi). Thus if only 90% of the cylinders are required to be better than 3500 psi, one would use the 3910 psi as the design base. Another curve shows how to get 95% better than 3500 psi (giving 4035 psi as the design base).

Unfortunately in both these cases it appears that some cylinders will fall short of reaching the 3400 psi requirement: the curve will have to move still more to the right. But it is evident we can never move the curve far enough to the right to assure that not a single cylinder will fall below 3400 psi. Our probability scale (like any log scale) pushes 0% down to infinity, and the 100% mark is pushed up an infinite distance as well. Some nonpassing cylinders *must* be permitted, or we must spend infinite dollars to achieve a rather pedestrian accomplishment. This is not unlike the environmental problem of achieving 100% pure water, 100% clean air, and 100% reduction of stack and auto exhaust emissions—all these goals would be achievable for an infinite price.

At all events, in our concrete cylinder problem it is likely that economic

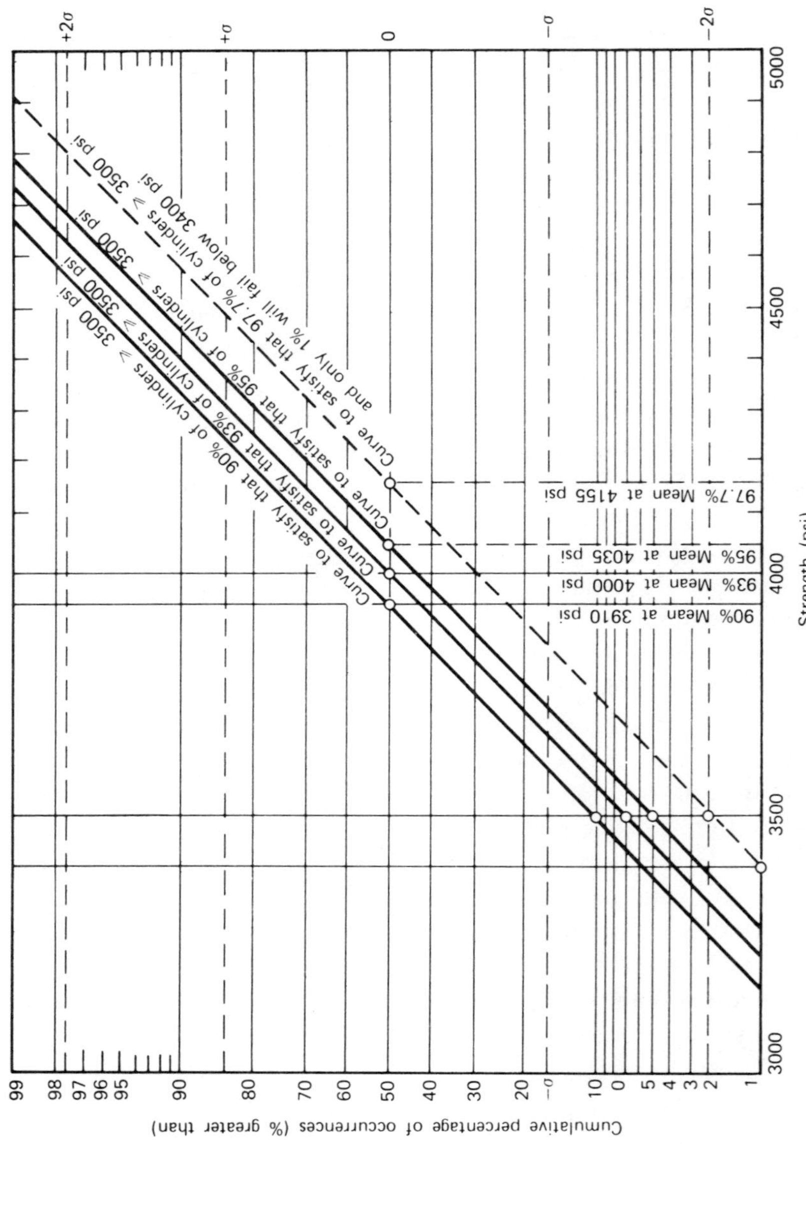

Fig 5.11 Normal distribution for concrete cylinder strengths, plotted on arithmetic probability paper.

reality will dictate that the final curve be considered as meeting the specification. Or, conversely, the specification may now be written to accommodate reality. It intersects the 3500 psi line at 2.3% (97.7% pass) and the 3400 psi line at 1%. This puts the mean test strength up to 4155 psi. It depicts the design that will meet the specification that 97.7% pass the 3500 psi test and only 1% fail to reach the 3400 psi mark. (Also see Section 6.4.)

In this particular example of concrete cylinders, the information on the "standard deviation" in strength was very definitely known to be "standard error" or "standard deviation of the mean" by reason of the statement on the reporting. Consequently, unlike the case of the light bulbs, one could not do any further analysis with a value less than that given for the sigma value. It should be remembered, however, that an investigation of the standard deviation of a "batch" of samples should leave open the possibility of prediction by use of the standard error (σ_m), and not just the standard deviation (σ_s).

Six

Reliability of Repeated Measurements

6.1 Multiple Measurements

From the foregoing treatment of the sigma error it is apparent that multiple measurements must be made of any quantity if anything is to be known about the precision (and, consequently, the accuracy) of the measurement. Frequently, careful individuals make a second or check measurement, a sort of "look-twice-and-be-sure" response. The question then arises: How many measurements must be made to achieve an acceptable level of precision?

Although it is sometimes specified that the standard deviation of a single value (σ_s) or the standard error of the mean (σ_m) shall not exceed a certain size, many times the acceptable level of precision is specified by fixing the number of times some observations must be made. In still other instances, the engineer must make his own determination of how precise to be and how many times to measure.

6.2 Number of Repeated Measurements

If we examine the equation for standard deviation, we see that it is essentially

$$\sigma_s = \sqrt{\frac{\Sigma v^2}{n-1}} = \frac{K}{\sqrt{n-1}}$$

where K may be assumed to remain fixed for the given series of measurements. It will not remain invariant, but if it did (for our purpose here), one can see in Fig. 6.1 that σ_s is not tending greatly to get any better (to decrease) beyond 10 or 15 measurements. It is justifiable on this basis (the study of the denominator) not to insist on repeating measurements beyond 15 or so to reduce the uncertainty of the mean. In practice this can be a fair guide, although some statisticians insist on sample sizes of 25 or larger.

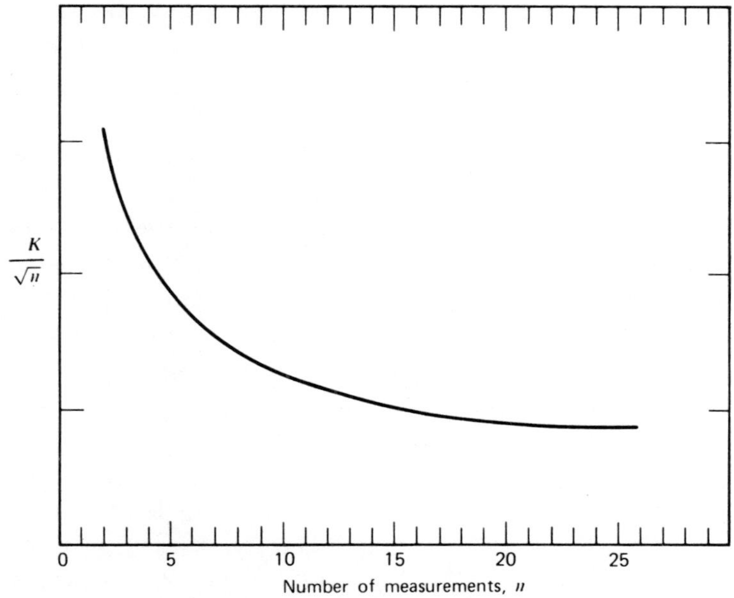

Fig 6.1 The effect of sample size.

6.3 Rejecting Measurements

Occasionally, when examining a set of measured values, some individual measurement is found to differ so widely from others in the same set that we suspect the discrepancy may be due to a mistake. If the widely divergent value is retained, it will produce a marked effect on the value of the arithmetic mean and its standard deviation, which represent the whole set of measurements. In such a case it may be well to exclude this measurement from the set.

The criterion and procedure for rejecting measurements are based on the principle that the range $\pm\sigma_s$ embraces 68.3% of the errors, $\pm 2\sigma_s$ embraces 95.5% of the errors, $\pm 3\sigma_s$ embraces 99.7% of the errors, and $\pm 3.29\sigma_s$ embraces about 99.9% of the errors. Hence the probability of making an error greater than $3.29\sigma_s$ is only about 1 in 1000, and we may safely assume that an apparent error of magnitude much greater than that must be a mistake.

Thus if a "wild" value is found in a set of measurements, compute the arithmetic mean by using the "wild" value, and find the σ_s. Next, discover $3.29\sigma_s$ and discard the suspected "wild" value(s) that lie beyond this amount. Then recompute the entire set, using only the good values. (Sometimes simple inspection initially will enable one to discard a "way out" reading.)

6.4 Maximum Possible Error

Consideration of the "maximum" error likely to occur in a set of measurements brings to light an extremely important reason for making multiple measurements of any given quantity. Such repeated measurements assure us that we are within a definite range of the "true" value. The maximum error ($\pm 3.29\sigma_s$) can be computed and used as a measure of the correctness of other check values obtained.

For example, in a given set of 20 measurements of a fluid flow in a pipe, the quantity (Q) is found to be 0.817 cubic foot per second (cfs) with a σ_s of ± 0.0025. By another single measurement the Q is determined to be 0.829 cfs, and the question of its validity is raised. The maximum error ($3.29\sigma_s$) likely to occur is ± 0.0082, causing us to have no confidence in any value outside the range 0.809–0.825. Thus we suspect that the value 0.829 contains a mistake, that it has not been corrected for some systematic error(s), or that the method used to determine it is invalid.

6.5 Use of Maximum Error

Throughout measurements, the "maximum" error ($3.29\sigma_s$) must be used as a criterion in any operation that requires "exactness." The placement of a bridge pier a half-mile out in open water by measurements from shore, for instance, must be sure and certain. The distance and angles used to compute the over-water lengths and establish the pier coordinates must have a known maximum error within an established tolerance. It will not suffice to know that we are "probably" (50–50 chance) correct, or even 68.3% certain. In this case, the $99.9^{+}\%$ is most likely required.

6.6 Significance

The foregoing concepts bring to light the notion of "significance" in statistical work. The term "significance" is used to indicate that the odds are heavy against the deviation from the expected value for something occurring by chance as a result of random sampling. In practice, odds of 19 to 1 against an occurrence by chance are taken as indicating that the occurrence is "significant." Some prefer heavier odds, such as 99 to 1, before conceding that some happening has significance; but in general, a probability of 19 to 1 is used as standard. The 19 to 1 corresponds roughly to the odds of getting a deviation from the mean of a normal distribution greater than twice the standard deviation, $2\sigma_s$ either positively or negatively.

6.7 Using the Standard Error to Compare Sets

The standard error (σ_m) is useful when comparing the mean values of two or more different sets of measurements. In the earlier example involving 20 measurements of an angle (Example 1 in Section 5.13), it is obvious that an entirely different 20 results might have been obtained, even by the same observer at the same time under identical conditions. We assume, knowing no better, that the 20 values recorded represent a good distribution, making a good bell-shaped curve. It may well be that the set of 20 is a trifle skewed to right or to left, however, and to some extent the mean will be affected. This was not checked by plotting to discover any skew, though this might be done.

Assume next that another set of 20 observations is made, this set having a different skew (perhaps) and a different mean. It is then valid to ask the following question: Are these two means significantly different? In other words, of all the hundreds and hundreds of possible observational values, the 40 here recorded are only two sets of random samples. And the two means of the two samples will be distributed with a standard error normally, meaning that each observed mean should be distributed about the "true" mean in a normal fashion.

6.8 Standard Error of the Difference Between Means

By computing the standard errors—one for each set—and comparing the two, we can discover whether their respective mean values differ significantly. We can do so simply by computing the standard error of the difference between the means by this formula (see Section 7.3):

$$\sigma_{\text{diff}} = \sqrt{[\sigma_m^2]_A + [\sigma_m^2]_B}$$

We then observe the agreement between the σ_{diff} and the actual difference between the mean values. If the actual difference between the two means is *greater than twice* this value (i.e., $d > 2\sigma_{\text{diff}}$), they are significantly different values. This indicates that it is unlikely (beyond a 19 to 1 chance) that the means represent the same measurement or the same measuring conditions.

Example 1. Two sets of measurements of angles are made, each consisting of 20 measurements. The mean and the standard error of each have been computed. Determine if these two means are significantly different.

Set	Mean value	n	σ_m	σ_m^2
A	19°27′36.12″	20	±0.42″	0.1764
B	19°27′34.92″	20	±0.67″	0.4489
Difference	01.20″			

The standard error of difference between A and B is:

$$\sigma_{\text{diff}} = \pm\sqrt{0.1764 + 0.4489}$$

$$= \pm\sqrt{0.6253} = \pm 0.79''$$

$$2\sigma_{\text{diff}} = \pm 1.58''$$

Comparison of this with the difference between the two means above (1.20″) indicates that these two sets do validly represent the value and that either mean can be used as "best" value of the quantity measured. (In fact, these two mean values could be averaged together to get a "best" value by a "weighted mean" computation, as Chapter 8 discusses.)

Example 2. A baseline is measured several times by each of two different parties under different climatic circumstances, with different (but standardized) tapes, but with methods of the same high order of precision. Comparison of results shows the following:

Set *A*	Set *B*
Value 1819.127 m	1819.210 m
$\sigma_m = 0.019$ m	$\sigma_m = 0.031$ m

σ_{diff}:

$$\sqrt{0.019^2 + 0.031^2} = \sqrt{0.001324}$$
$$= \pm 0.036$$

$2\sigma_{\text{diff}}$:

$$\pm 0.072 \text{ m}$$
$$1819.210 - 1819.127 = 0.083 \text{ m}$$

It should be concluded that there does exist a significant difference between the two results, being more than a chance deviation (expected 1 in 20 times). A difference of more than $2\sigma_{\text{diff}}$ occurs too infrequently by chance (i.e., by random fluctuation) to be written off as caused by random errors.

6.9 Limitation of Standard Error of the Difference

Note here that strictly speaking the formula for σ_{diff} cannot be validly used for comparison purposes unless there are a large number of measurements, say about 50 in all. Some statisticians insist on this. For most practical purposes, however, this formula *can* be used, with proper caution, as an indicator of sameness if fewer than 50 measurements comprise each set.

Seven

Propagation of Errors in Computing

7.1 Introduction

When using direct measurement values to compute final results, such as to reach indirect measurements, it is necessary to guard against carrying excessive random error into the result. Thus it is necessary to know the size of the error in the result after arithmetic operations have been performed on any measured value or on the mean of several values. The general rules of error propagation refer to σ_s, σ_m, E_p, or any similar random errors. (In the following operations, the error of the mean is generally used, though the subscript m is omitted for clarity.)

7.2 Addition of Values Containing Errors

According to the laws of probability, when quantities are added, each containing an error (σ, E_p, 2σ, or any similar accidental type), their sum contains an error equal to the square root of the sum of the squares of the errors of the added quantities. Thus

$$E_{\text{sum}} = \sqrt{E_1^2 + E_2^2 + E_3^2 + \cdots + E_n^2}$$

For example, if line $ABCD$ is comprised of three measured segments, each value given being the mean of many measurements:

$$AB = 107.162 \pm 0.002 \text{ cm}$$
$$BC = 491.043 \pm 0.010 \text{ cm}$$
$$CD = 216.191 \pm 0.005 \text{ cm}$$

$$AD = 814.396 \pm \sqrt{0.002^2 + 0.010^2 + 0.005^2}$$
$$= 814.396 \pm 0.011 \text{ cm}$$

This is a valid rule to follow in sums of quantities affected by errors, but some authorities insist that the *maximum* error in the sum of the three quantities is $\pm(0.002 + 0.010 + 0.005) = \pm 0.017$ cm. They say (safely enough) that "the error in the sum is not greater than the sum of the errors in the added quantities."

Of course the validity of this "maximum" error is unquestionable, but it is not reasonable to insist on such certitude because each of the three segment values is the mean of several readings, and each is labeled with an error value signifying a reasonable uncertainty. It is very unlikely that each of the quantities is uncertain by the maximum amount shown *and* each in the same direction at the same time, *all* being either too small or too large. More likely, the measured quantities differ from their true values by less than the indicated value of the uncertainties, with some plus and others minus. Thus the possible error in a sum (or other calculated quantity) is much more logically represented by a probable than by an absolute value. Hence we use the generally accepted probability-type square root computation for the error in a sum.

7.3 Subtraction of Values Containing Errors

If we agree that subtraction is the addition of one positive and one negative value, the error of the result can be seen to be validly computed in the same manner as the addition error:

$$E_{\text{diff}} = \pm\sqrt{E_1^2 + E_2^2}$$

For example, in subtraction of two angular values:

$$\begin{array}{r} \text{Angle } AOC = 87°45'15'' \pm 05'' \\ -\text{Angle } BOC = 41°53'50'' \pm 10'' \\ \hline \text{Angle } AOB = 45°51'25'' \pm \sqrt{05^2 + 10^2} = 11.2'' \end{array}$$

7.4 Multiplication of Values Containing Errors

In multiplying two quantities that contain errors, this is the general form:

$$(A \pm E_A)(B \pm E_B) = AB \pm E_A B \pm E_B A \pm E_A E_B$$

To neglect the last term is not serious, since both factors are extremely small. Hence it may be seen (from the previous rule for addition of errors) that

$$E_{\text{product}} = \pm\sqrt{(E_A B)^2 + (E_B A)^2} = AB\sqrt{\frac{(E_A B)^2 + (E_B A)^2}{A^2 B^2}}$$

$$= \pm AB\sqrt{\left(\frac{E_A}{A}\right)^2 + \left(\frac{E_B}{B}\right)^2} \qquad \text{(general form)}$$

Example. Compute the area of the measured field and the error of the area, given these values:

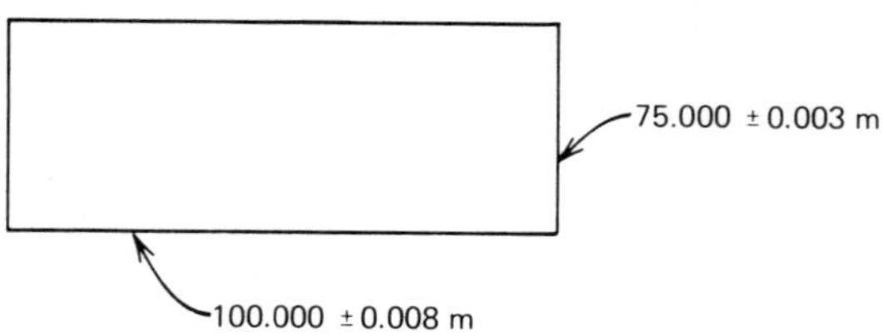

Thus the area (A) of the rectangle shown is given by

$$A = 7{,}500.00 \pm \text{error}$$

$$\text{error} = \pm\sqrt{(100 \times 0.003)^2 + (75.0 \times 0.008)^2}$$

$$= \pm\sqrt{0.0900 + 0.3600} = \pm\sqrt{0.4500} = \pm 0.67\ m^2$$

The best value is 7,500.00 ± 0.67 m^2.

Similarly, the product of several factors has this error:

$$E_{\text{product}} = \pm ABC \cdots N \sqrt{\left(\frac{E_A}{A}\right)^2 + \left(\frac{E_B}{B}\right)^2 + \left(\frac{E_C}{C}\right)^2 + \cdots + \left(\frac{E_N}{N}\right)^2}$$

(general form)

Each fraction under the radical may be called the "relative error" of the designated factor, or the percentage error.

7.5 Division of Values Containing Errors

Though the derivation is beyond the scope of this treatment, the error in a quotient $(A \pm E_A) \div (B \pm E_B)$ is

$$E_{\text{quotient}} = \frac{A}{B} \sqrt{\left(\frac{E_A}{A}\right)^2 + \left(\frac{E_B}{B}\right)^2}$$

or, written in another form:

$$E_{\text{quotient}} = \frac{1}{B^2} \sqrt{(E_A B)^2 + (E_B A)^2}$$

Thus in a given case where it is desired to stake out a rectangular plot *PQRS* for a building whose area must be 200.000 ± 0.165 ft^2 and whose one fixed dimension necessarily is 10.000 ± 0.008 ft, the perpendicular dimension is

$$\begin{aligned} PQ &= \frac{200.000}{10.000} \pm E_{\text{quotient}} \\ &= 20.000 \pm \frac{200.000}{10.000} \sqrt{\left(\frac{0.165}{200.000}\right)^2 + \left(\frac{0.008}{10.000}\right)^2} \\ &= 20.000 \pm 20.000 \sqrt{68 \times 10^{-8} + 64 \times 10^{-8}} \\ &= 20.000 \pm 20.000(11.5 \times 10^{-4}) \\ &= 20.000 \pm 0.023 \text{ ft} \end{aligned}$$

7.6 Multiplication of Values Containing Errors by a Constant

We often encounter the conversion of a measured value into other units, or the totaling of the measured value a number of times. The value of the product $C(A \pm E_A)$ is seen to be $CA \pm CE_A$, and

$$E_{\text{product}} = \pm CE_A$$

Hence, for example, if a length is measured as 3612.28 ft ±0.12, and to compute another result it is multiplied by

$$\frac{\sin 90°00'00''}{\sin 39°51'32''} = \frac{1.00000000}{0.64089901} = 1.56030824$$

$$\text{the product} = 5636.270 \pm 1.5603 \times 0.12$$
$$= 5636.270 \pm 0.187 \text{ ft}$$

Similarly, to convert 802.316±0.027 m to feet, the proper value in feet is:

$$\text{product} = 802.316(3.2808399) \pm 0.027(3.2808)$$
$$= 2632.270 \pm 0.089 \text{ ft}$$

(The product here, incidentally, may properly be given to seven significant figures, since the 8 in 802.316 is almost *two* significant figures, being very nearly 10.)

The preceding conversion used the new (1959) relationship of feet and meters, based on 1 in. = 2.54 cm exactly. To convert this value to the U.S. Survey Foot (based on 39.37 in. = 1 m exactly), the conversion would be:

$$\text{product} = 802.316(3.2808333) \pm 0.027(3.2808)$$
$$= 2632.265 \pm 0.089 \text{ ft}$$

7.7 Elevation to a Power of a Quantity Containing an Error

If we regard the raising to the nth power to be simply a multiplying of a quantity by itself n times $(A + E_A)^n$, then the error of the result would seem to be, as ascertained from the general form for a product (see Section 7.4):

$$E_{\text{power}} = A^n \sqrt{n\left(\frac{E_A}{A}\right)^2} = \frac{E_A}{A} A^n \sqrt{n}$$
$$= E_A A^{n-1} \sqrt{n}$$

Our assumption would imply, however, that the E_A value is not fixed for the quantity A but may differ in value each time A is used. Such is not the case, since the quantity A has a fixed value and a fixed E_A throughout. Once the measurement of A has been completed, the error E_A is fixed and will not vary so as to compensate or to tend to compensate. Hence the correct evaluation can be made of E_{power} if we note that error E_A will be involved n times and not $\sqrt{n}$ times. Thus correctly:

$$E_{\text{power}} = nA^n\sqrt{\left[\frac{E_A}{A}\right]^2} = nA^{n-1}E_A$$

Example. If a perfect cube of silicon crystal is measured by a micrometer microscope to be 4.00 ± 0.02 micron, its volume is

$$64.0 \pm (3 \times 4.00^2 \times 0.02) = 64.0 \pm 0.96 \text{ micron}^3$$

The implication is that the cube is measured along one edge only, and the assumption of perfect cubical shape is made.

Example. Find the error in the volume of a sphere whose diameter is repeatedly measured and is found to be 4.000 ± 0.006 in. Let

$$A = \text{radius} = 2.000$$

$$E_A = \pm 0.003$$

$$\text{volume of the sphere} = \frac{4\pi}{3}(2.000)^3$$

$$= 33.510 \text{ in.}^3$$

$$\text{error of volume} = \frac{4\pi}{3}\left[3(2.000)^2(0.003)\right]$$

$$= \pm 0.048\pi$$

$$= \pm 0.151 \text{ in.}^3$$

7.8 Root of a Quantity Containing an Error

If we consider the problem to be $(A \pm E_A)^{1/n}$, by using the preceding equation:

$$E_{\text{root}} = \frac{1}{n} E_A A^{1/n-1} = \frac{E_A \sqrt[n]{A}}{nA}$$

Thus if a true square must be measured out such that its area is 16.000± 0.008, the length of each side must be

$$L = \sqrt{16.000} \pm \frac{0.008\sqrt{16.000}}{2 \times 16.000}$$

$$= 4.000 \pm 0.001$$

7.9 Other Operations

The foregoing principles are basic in the study of calculations that involve quantities containing errors, but certain frequently used concepts should be restated here for emphasis.

1. The error of the mean of a series of measurements is found, from principles of Sections 7.2 and 7.5, thus:

$$\text{mean} = \frac{(A_1 + A_2 + A_3 + \cdots + A_n) \pm \sqrt{v_1^2 + v_2^2 + v_3^2 + \cdots + v_n^2}}{n}$$

$$= A_{\text{mean}} \pm \frac{\sqrt{\Sigma v^2}}{n}$$

   Note that $\sqrt{\Sigma v^2}\,/n$ is the *mean variation* or *mean deviation* (Section 5.7). But note that it is very nearly the standard deviation divided by $\sqrt{n}$, thus:

$$\text{error of mean} = \frac{\sqrt{\Sigma v^2}}{n} = \sqrt{\frac{\Sigma v^2}{n}}\sqrt{\frac{1}{n}} \cong \frac{\sigma_s}{\sqrt{n}} = \sigma_m$$

   Hence the error of the mean of the series is quite nearly the *standard error* for the series (Section 4.5).

2. The addition of quantities having the same size of errors (see Section 7.2) can be represented thus:

$$(A \pm E_1) + (B \pm E_1) + (C \pm E_1) + \cdots + (N \pm E_1)$$

$$= (A + B + C + \cdots + N) \pm \sqrt{E_1^2 + E_1^2 + E_1^2 + \cdots + E_n^2}$$

$$= (A + B + C + \cdots + N) \pm E_1\sqrt{n}$$

For example, if a 300-ft tape has been corrected for all its systematic errors but has (necessarily) some accidental errors affecting it and its use, and the net accidental error (E) for each tape length is ± 0.024 ft, what is the error likely in a distance of 1545.0 ft? Since $1545/300 = 5.030$ tape lengths,

$$E_{1545} = \pm 0.024\sqrt{5.030} = \pm 0.054 \text{ ft}$$

Eight

Errors and Weights

8.1 Weight and Reliability

A further and more common use of the standard deviation (σ_s) or the standard error (σ_m) is to assign weights to measurement values that are to be added, averaged, or otherwise used in computation.

To place some sort of confidence tag on the mean of any sample, we usually employ the standard deviation (σ_s); but if the sample is large, we can justifiably use the standard error (σ_m). Weights are assigned with some regard to the certainty of each set of measurements (or to each measurement) being averaged, and the certainties of these are discovered from their σ_s or their σ_m. Weights distinguish which are the more reliable values and how much more reliable these are. Assigning a weight to each measurement or set will allow each to exert its proper influence in computations based on measurements.

8.2 The Weighted Mean

The "average" of the means of several sets of measurements that will give correct cognizance to the influence each should play in determining the "true" value is known as the *weighted mean*. It is the only logical key to combining means of several sets, and it is computed by simply multiplying each measurement value x by its properly assigned weight factor w, then

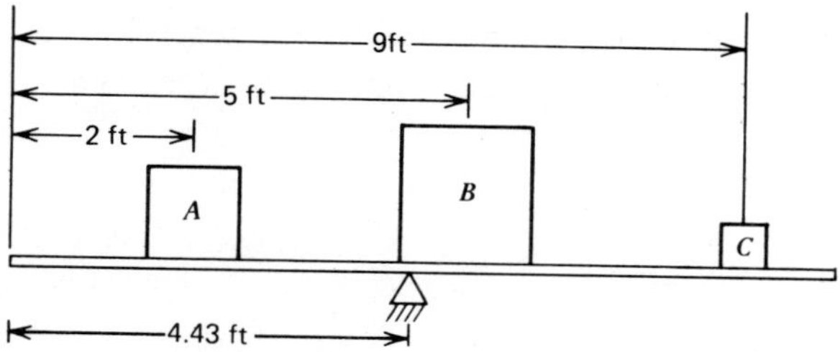

Fig 8.1 First moment principle.

dividing the sum of the products by the sum of the weights:

$$\bar{X}\,(\text{mean}) = \frac{\Sigma(xw)}{\Sigma(w)}$$

This is analogous to the "first moment" principle, as exemplified by the problem of finding the location of the center of gravity of three cubes on a board (where volume is a "weight factor"); see Fig. 8.1.

Cube	Volume	x	xV
A	4	2	8
B	9	5	45
C	1	9	9
$\Sigma =$	14		62

Mean:

$$\bar{x} = \frac{\Sigma xV}{\Sigma V} = \frac{62}{14} = 4.43$$

This indicates the board would balance if a pivot were placed at 4.43 ft, the center of gravity (centroid) of all the volumes.

8.3 Assigning Weights

In determining the weights to be assigned to measured quantities, various guides are used as reasonable indicators, ranging from simple judgment to "exact" statistical computation. Besides certain general rules for assigning weights, we should consider three simple notions, which are commonly used.

8.3.1 Use of Judgment

Sometimes the judgment of the measurer must play a major role in assigning importance to measurements. For example, we may have ascertained the velocity of the maximum wind in a hurricane by using two different instruments stationed at $1\frac{1}{2}$ miles apart, but later one of the instruments was blown over by the wind. The observer may remember doubting the value of its readings and, because the destroyed instrument is now unavailable for testing, he would correctly discount much of the worth of this recording. It may be his judgment that only one-tenth as much weight should be accorded to its reading, thus:

Instrument	Maximum, V (mph) velocity	Weight, w	wV
A	87	10	870
B (destroyed)	81	1	81
$\Sigma =$		11	951

Estimated maximum velocity (mean):

$$\frac{951}{11} = 86\tfrac{1}{2} \text{ mph}$$

Often in engineering, judgment is also required in locating a factory or other installation when two or more sites may offer varying degrees of similar advantages. Price of land, labor market, transportation availability, raw material availability, nearness of product market, weather, and so on, are accorded arbitrary or rational weight factors for each site to facilitate the comparison.

8.3.2 *Use of Indexes*

Often devices are available for assigning weights, as illustrated in the following comparison that uses the "index" of the dollar to assign weights for a cost problem.

year	Cost of typewriter c	Index of the dollar, (w) (1940 = 1.00)	Comparable 1940 costs, c/w	Comparable 1960 costs $\frac{c}{w} \times \frac{1.82}{1.00}$	Costs x worth cw
1940	\$115	1.00	\$115	\$209	\$115
1950	\$157	1.49	\$105	\$192	\$234
1960	\$203	1.82	\$112	\$203	\$369
$\Sigma =$		4.31			\$718

Here it is seen that the dollar "index" (compare "cost-of-living index") is used as a weight factor to compare costs. The weighted mean cost of a typewriter during two decades (for whatever good it might do) is also:

$$\bar{c} = \frac{\$718}{4.31} = \$167$$

although this is not really in terms of any one year's dollar and serves no real purpose in a comparison. (The valuable comparable costs occur in the tabulation.)

8.3.3 *Concept of Numbers*

Another weighting device is the concept of numbers. For instance, if three clocks say 3 o'clock, we reason that it is more reliably 3 o'clock than if only one clock said so. Or if five people say that such and such an event happened, it gains credence. Note that these should be, in all such cases, independent observations. We have three independent judges at the end of a race, for instance, and three independent timers. The three "times" of the race may be given equal weight because the people are considered to be equally responsive, and all other conditions are approximately equivalent.

GENERAL RULES FOR ASSIGNING WEIGHTS

8.4 Equal Weights

When the same conditions prevail, equal weights can be assigned to two measurements of the same quantity, or to two sets of measurements of the same quantity.

Example. A distance *PQ* was measured six times by team *A* and six times by team *B*, each team using identical equipment and care. Their results must be accorded equal weight, thus:

Team	Value, X (m)	Weight, w	wX
A	15.6172	1	15.6172
B	15.6184	1	15.6184
$\Sigma=$		2	31.2356

Mean:

$$\bar{X}=\frac{31.2356}{2}=15.6178 \text{ m}$$

This is, of course, to take the simple arithmetic mean, which is a special case of the weighted mean (when weights are equal).

A variation of this equal-weight principle is illustrated by adding the means of two measured values (achieved under identical conditions) to get a sum. For example, angle *AOB* (30°13′42″) is added to angle *BOC* (15°49′50″) to give angle *AOC* (46°03′32″). There is nothing else that can be done, for we do not have a check value for the sum. This becomes clearer in Section 8.5.

8.5 Equal Weights, with Exact Check Value

In a given instance where a sum of three equal-weight mean values must equal a fixed value, adjustment can be made equally to each mean so that the sum of the parts exactly equals the whole. For example, if each of the three angles of a triangle is measured several times, the mean values, when added, must equal (or else be adjusted to) 180° exactly. Here is such a situation.

Angle	Observed mean value	Weight	Adjustment (02.7″ ÷ 3)	Adjusted values
A	49°46′45.7″	1	+00.9″	49°46′46.6″
B	34°51′39.8″	1	+00.9″	34°51′40.7″
C	95°21′31.8″	1	+00.9″	95°21′32.7″
Σ=	179°59′57.3″ (deficiency = 02.7″)		(02.7″)	(180°00′00.0″)

Note that the exactly fixed value (180°00′00.0″) can be considered to have "infinite" weight and thus zero adjustment.

Similar situations arise when angular values of any polygon are measured, when angles around a point are measured to "close the horizon," and in other such cases. However measurements that can be checked against exact values are not very frequent.

8.6 Equal Weights, with Arbitrary Check Value

It may happen, however, that a "total" or a check value may be fixed merely for convenience, or to preclude continuing endless minor adjustments of its value. For example, a property line fixed by monuments, set once as 891.750 ft long, is now measured in two segments that do not total exactly that distance. Adjustment is made of the two segments, once it is seen that they total very nearly that fixed value. The mode of adjustment is that of "prorating" the discrepancy on the basis of segment lengths, assuming that any errors are proportional to the segment lengths. (The reason for the method of prorating does not follow from the present treatment, but comes instead from an understanding of how errors are made in taping.)

Segment	Value	Weight	Adjustment	Adjusted value
1	216.502	1	$\frac{216.5}{891.8}\times 0.011=0.0027$	216.499
2	675.259	1	$\frac{675.3}{891.8}\times 0.011=0.0083$	675.251
Σ=	891.761		(0.011)	(891.750)
	Fixed 891.750 (discrepancy of 0.011)			

8.7 Equal Weights of Components and Total

More frequently, instead of an exact total against which to check, the only available value for a total is a measured value. This is equivalent to a case of three equally weighted partial measurements and a total—four weighted measurements—with adjustment to be made to the three partials and to the total. Values tabulated are lengths, segments of the line *AD*.

Segment	Value (*m*)	Weight	Adjustment	Adjusted value (m)
AB	15.4179	1	+0.00553	15.4234
BC	8.0124	1	+0.00287	8.0153
CD	20.6008	1	+0.00738	20.6082
Σ=	(44.0311)			(44.0469)
AD	44.0627	1	−0.01577	44.0469
Difference	0.0316			0.0000

Prorating here can be done best by considering that the measured distance is twice the total length (i.e., a round trip). The adjustments are, for the tabulation above:

$$\frac{15.4179}{44.0627+44.0311}\times 0.0316=0.00553 \qquad \text{(to be added, clearly)}$$

$$\frac{8.0124}{44.0627+44.0311}\times 0.0316=0.00287 \qquad \text{(also to be added)}$$

and so on, including that for *AD* (which must be subtracted).

8.8 Weights Proportional to Number of Measurements

When conditions are identical except that one quantity is measured more times than another, weights are reasonably assigned in proportion to the number of measurements. Thus length Peabody–Capstone Head was measured by Loran five times in one set and three times in another, with these resulting means:

Set	Value, (km)	Measurements	w	Weight, wX
1	187.234	5	1.667	312.056
2	187.160	3	1.	187.160
$\Sigma =$			2.667	499.216

Weighted mean:

$$\bar{X} = \frac{499.216}{2.667} = 187.206 \text{ km}$$

This notion is not to be confused with that used in computing the arithmetic mean from a frequency table (Section 4.8). The assigning of weights in this instance has nothing to do with the magnitude of the values found; it is related only to the effort expended.

8.9 Weights and Errors

Because the standard error (σ_m) of a set of measurements signifies the degree of uncertainty (and, therefore, certainty), it furnishes a guide to the weight we may accord to the mean of a set. A large standard error, induced by large dispersion and, mainly, by too few measurements in a set, indicates that the set should be regarded as less certain than a set with a small standard error. Therefore, the set with the larger standard error should be accorded less weight, and the set with the smaller error more weight.

8.10 Weights and the Standard Error

From the definition of standard error

$$\sigma_m = \frac{\sigma_s}{\sqrt{n}}$$

it can be seen that the error varies inversely with the square root of the number of observations or, stated differently, the number of observations varies inversely with the square of the standard error. Because weights ought to be assigned to sets in proportion to the number of measurements in the sets, weights ought to be varied according to the inverse square of the standard errors of the sets, or

$$W_1 : W_2 : W_3 : \ldots = \frac{1}{(\sigma_{m_1})^2} : \frac{1}{(\sigma_{m_2})^2} : \frac{1}{(\sigma_{m_3})^2} : \ldots$$

If, for example, the stress in the steel wall of a high-pressure cylinder under load is recorded in four sets of measurements by strain gauges, each set being of different uncertainty, the weighted mean can be found.

Set	Stress, X (psi)	σ_m	Weight ratio	Weight factor, w	wX
A	39,765	± 125	$(1/125)^2 = 6.40 \times 10^{-5}$	1.5376	61,142.7
B	39,810	± 80	$(1/80)^2 = 1.5625 \times 10^{-4}$	3.7539	149,442.8
C	39,716	± 155	$(1/155)^2 = 4.1623 \times 10^{-5}$	1.0000	39,716.0
D	39,791	± 140	$(1/140)^2 = 5.1020 \times 10^{-5}$	1.2258	48,775.8
$\Sigma =$				7.5173	299,077.2

Weighted mean:

$$\bar{X} = \frac{299{,}077.2}{7.5173} = 39{,}785.2 \text{ psi}$$

Note that the value with the smallest standard error (σ_m) exerts the greatest influence in fixing the value of the weighted mean.

8.11 Weights and Adjustments

Since weights signify the degree of reliability of measurement sets, it is apparent that a value with a high weight factor is not to be greatly adjusted (some would say "corrected") and, conversely, a value with a low weight factor should be adjusted more. Therefore the adjustments ("corrections," loosely) should be made in inverse ratio to the weights.

$$\frac{c_1}{1/W_1} = \frac{c_2}{1/W_2} = \frac{c_3}{1/W_3} = \frac{c_4}{1/W_4}, \ldots$$

where c means "correction."

But we have already seen that the weights vary inversely with the square of the standard error. Hence

$$\frac{c_1}{(\sigma_{m_1})^2} = \frac{c_2}{(\sigma_{m_2})^2} = \frac{c_3}{(\sigma_{m_3})^2} = \frac{c_4}{(\sigma_{m_4})^2} = \cdots$$

or, stated another way, the adjustments ("corrections") vary *directly* as the square of the standard errors.

In the case of the triangle of Section 8.5, let us suppose the weights were as shown in the accompanying table. The computation would be:

Angle	Mean value	Weight, w	Adjustment ratio	Adjustment*	Adjusted value
A	49°46′45.7″	4	1/4=0.25	+00.4″	49°46′46.1″
B	34°51′39.8″	1	1/1=1.00	+01.5″	34°51′41.3″
C	95°21′31.8″	2	1/2=0.50	+00.8″	95°21′32.6″
$\Sigma=$	179°59′57.3″		1.75	+02.7″	180°00′00.0″

*Adjustment: total=02.7″. Note that the adjustment is made in *inverse* ratio to the weights.

for A: $\dfrac{0.25}{1.75} \times 02.7'' = 00.4''$

for B: $\dfrac{1.00}{1.75} \times 02.7'' = 01.5''$

for C: $\dfrac{0.50}{1.75} \times 02.7'' = 00.8''$

8.12 Unequal Weights with Exact Check Value

As indicated in Section 8.5, the availability of an exact check value allows the adjustment of means of measurement sets so as to fit the exact condition, as the following example illustrates.

Example. At a given station, a transit was used to measure four angles, closing the horizon. The value for each angle was found several times, and a σ_m for each angle was found, as the table shows. Adjust the values by properly weighting them, in accordance with the principles of Section 8.11.

Note that in each case of this adjustment, the *weights* are not used. Rather, the adjustment factors are obtained from the standard errors immediately. In other words, adjustments may be made in direct proportion to the square of the standard errors.

Angle	Value	σ_m	Adjustment ratio	Adjustment*	Adjusted values
AOB	34°15′31″	±03″	$3^2 = 9$	−00.6″	34°15′30.4″
BOC	23°49′55″	±10″	$10^2 = 100$	−06.5″	23°49′48.5″
COD	190°50′54″	±05″	$5^2 = 25$	−01.6″	190°50′52.4″
DOA	111°03′49″	±02″	$2^2 = 4$	−00.3″	111°03′48.7″
Σ=	360°00′09″		138	−09.0″	360°00′00.0″

*Adjustment: total = 09.0″.

for *AOB*: $\frac{9}{138} \times 09 = 00.6''$

for *BOC*: $\frac{100}{138} \times 09 = 06.5'', \ldots.$

This is equivalent to adjusting the values in *inverse* ratio to their weights (and their weights vary *inversely* with the square of their standard errors).

8.13 Standard Error of the Mean of Several Means

After finding the mean of values, it is important that a label, the standard error of the mean, be affixed to it. Similarly, one should wish to label the mean of several means of sets or samples with a single standard error. The use of the same weights will be used; the resultant standard error will reflect the several standard errors according to the weights used to find the mean itself. We employ the notion of Section 7.9. The calculation that follows is for the high-pressure cylinder problem of Section 8.10.

$$\sigma_m = \frac{1}{n}\sqrt{(\sigma_{m_1})^2 + (\sigma_{m_2})^2 + \cdots + (\sigma_{m_n})^2}$$

Set	σ_m	σ_m^2	Weight, w
A	±125	15,625	1.5376
B	± 80	6,400	3.7539
C	±155	24,025	1.0000
D	±140	19,600	1.2258
Σ=		65,650	7.5173

Standard error:

$$\sigma_m = \frac{\sqrt{\Sigma(\sigma_m)^2}}{\Sigma(w)} = \frac{\sqrt{65{,}650}}{7.5173} = \pm 34.1 \text{ psi}$$

Thus the mean pressure with its proper label is 39,785.2 ± 34.1 psi.

8.14 Illustrative Calculations

The following illustrative examples show something of the worth of the foregoing principles in practical application.

Example 1. A party did a differential level (tilting-type) several times from bench mark 39 to bench mark 40, using careful techniques, and they computed a mean result and σ_s. Later, using a less precise instrument and fewer runs, they ran back from 40 to 39 with a mean result and σ_s. Find the weighted mean of the values.

Set	Value, X (m)	σ_s	Weight ratio	Weight factor, w	wX
A	4.6123	±0.0007	$(1/7)^2=2041$	2.939	8.817
B	4.6128	±0.0012	$(1/12)^2=694$	1.000	8.000
$\Sigma=$				3.939	16.817

Weighted mean:

$$\bar{X} = 4.6120 + \frac{16.817}{3.939} = 4.6120 + 0.00043 = 4.61243 \text{ m}$$

Note the shortcuts: (*a*) weight ratios use only the significant figures, (*b*) in the last column, only the last digit of the number in the value column is multiplied by the weight factor. These keep the operations simple enough for slide rule calculation. Analysis of significant figures will demonstrate the validity of such shortcutting.

Example 2. Consider the following measurements of a peak in Colorado:

By plane table:	13,992	±18 ft
	13,996	±18 ft
By spirit leveling:	14,002	±4 ft
By theodolite (triangulation):	14,001	±2.5 ft
	14,003	±2 ft
	14,001	±2.5 ft
By Airphoto triangulation:	14,003	±3 ft
	13,999	±3 ft
	14.001	±4 ft

Calculate the mean elevation and also its range of error.

The solution, by the weighted mean method, requires the use of the tabulation that follows.

	X	E	Weight ratio, $1/E^2$	Weight factor, w	wX
1	13,992	±18	0.00309	1.0000	13,992.0
2	13,996	±12	0.00694	2.2481	31,464.9
3	14,002	±4	0.06250	20.2462	283,487.2
4	14,001	±2.5	0.16000	51.8302	725,675.4
5	14,003	±2	0.25000	80.9848	1,134,029.8
6	14,001	±2.5	0.16000	51.8302	725,675.4
7	14,003	±3	0.11111	35.9932	504,013.2
8	13,999	±3	0.11111	35.9932	503,869.3
9	14,001	±4	0.06250	20.2462	283,467.0
				300.3721	4,205,674.2

Weighted mean:

$$\bar{X} = \frac{\Sigma(wX)}{\Sigma(w)} + \frac{4{,}205{,}674.2}{300.3721} = 14{,}001.6 \text{ ft}$$

To calculate the error of the mean, one uses the notion

$$E_{\text{sum}} = \sqrt{E_a^2 + E_b^2 + \cdots + E_n^2}$$

and the notion that follows: any value (A) with its error (E) divided by a constant (k) gives a quotient

$$\frac{A \pm E}{k} = \frac{A}{k} \pm \frac{E}{k}$$

	E	w	$w(E^2)$
1	±18	1.000	324
2	±12	2.2481	324
3	±4	20.2462	324
4	±2.5	51.8302	324
5	±2	80.9848	324
6	±2.5	51.8302	324
7	±3	35.9932	324
8	±3	35.9932	324
9	±4	20.2462	324
		300.3721	2916

Error of the "sum" of all 300.3721 measured quantities:

$$E = \sqrt{2916} = \pm 54.0$$

Error of the *mean*:

$$E_m = \frac{\pm 54.0}{300.37} = \pm 0.18 \text{ ft}$$

Nine

Practical Application of the Theory of Errors in Measurement

9.1 Remarks

We have discovered that to be absolutely sure and certain of a measurement, we must repeat it nearly an infinite number of times (Chapter 3). But since we do not have a thousand years to live, we must use a more economical method of making measurements. It was subsequently indicated (Section 6.2) that we can generally be fairly certain of our precision if we measure 10, 15, or 20 times. But often enough we can still not afford even this smaller expenditure of effort. Obviously, then, a more practical approach must be found.

9.2 Standard Deviation a Criterion

The clue to the solution of this number-of-measurements problem lies in the analysis used to find the standard deviation itself. Section 3.8 noted that when measurements of a quantity are made in a given set, they must be all done under the same (standard) circumstances, by the same (standard) method if the results are to be comparable. In a word, standard deviation (σ_s) applies only to identically made measurements.

With this in mind, it now becomes apparent that we must standardize on a measurement method that will give us a set of measurements that have a certain fixed standard error (σ_m) of a size that we can tolerate. This, of course, would imply a method that will furnish a certain standard deviation (σ_s) for an individual measurement. The first step is to determine the acceptable maximum size of standard deviation.

9.3 Fixing on a Maximum Desired Error

In the selection of the largest size of error that can be tolerated, we must view the measurement's place in the entire project. Its contribution to the computation must be examined to establish its influence on the final result.

Selecting too large a tolerable spread (too large a standard deviation) might give too much leeway for sloppy work in making measurements, although it is frequently desirable that a measurement be made only roughly. Such work would probably by characterized as "rough" or of "low-order precision" in this case, rather than "sloppy," because essentially it would be in keeping with the low precision demanded.

On the other hand, when a very tight spread is required, for very precise work, it often is very difficult and expensive to meet this requirement. Here an economic balance must be struck in making demands for "first-quality" or "high-order" measurement.

9.4 Selecting a "Maximum Error"

Once the possible effect of the measured quantity is established, however, its tolerance can be fixed. The tolerance, limits of plus and minus error about a mean value, is effectively a precision of the measurements that are to be made. The tolerance says, in effect, that any single measured quantity must lie between certain values. If we wish to fix this at a maximum allowable error, it will mean that the limits of "maximum" error (3.29σ) can be set, giving us 100% certainty of our measurement. (See Table 5.1.)

It is possible, though, that a less rigid "maximum" or acceptable error might be used, say, 3σ or 2σ or the "90% error" (1.64σ). This frequently makes for considerably less effort, though at a risk that some 1 or 5 or 10% of the measurements will have greater errors than the prescribed maximum. Here again we must balance this risk against the economics of the given situation. Many times it is entirely admissible to use a lesser degree of certainty, especially if additional precautions (e.g., check measurements by another procedure) are employed.

9.5 Procedure for Limiting the Error

Let us reexamine the limits or error we should consider satisfactory, to be able to ascertain a method of achieving the security of knowing that our measurements will not exceed such an arbitrarily fixed limiting error. Suppose, for example, we wish to measure an angle and be certain that it is not more than $\pm 05''$ from the "truth." We could go through our infinite-number-of-measurments routine, or we could simply use a routine for measurement that is known to produce this size of standard deviation (σ_s), this size of "maximum" error (3.29σ), or this size of 90% error (1.64σ), and so on. If such a fixed routine (e.g., using a 10-second repeating theodolite and measuring twice or four times) has already been established, we can utilize this routine and then reasonably believe that our result will be within the limits desired.

9.6 Standardizing the Procedure

After having fixed the precision to be demanded in the measurement, we must examine procedural methods and fix on one that will give the desired result. If the measurement (or one similar to it) has never before been made, it will be necessary to make very many measurements of the same kind to establish the validity of the procedure and to enable a study of the precision obtainable. Such a case occurred when the speed of light was ascertained by use of a "measured mile" some years ago.

The ingredients of the method must be scrutinized for elimination of all but random errors; then the results must be analyzed for scatter, standard deviation, and so on, to discover the precision that can be expected from the method employed. Then the procedure can be fairly well characterized as being precise to $\pm\sigma_s$ (68.3% of the time) or $\pm 2\sigma_s$ (95.5% of the time), and so on.

9.7 Utilizing a Standard Procedure

Once the precision of the method or procedure is fixed by an "infinity" of measurements, however, and once the details of procedure are spelled out, it is apparent that when the method is used and all its prescribed details are adhered to, the results will be similar. In other words, a careful execution of a standard procedure enables us to rely on a whole family of measurements made by this procedure. We thus utilize the results of many hundreds of measurements made similarly at other times and other places, other quantities ascertained by other people, to warrant or guarantee the precision (hence the accuracy) of our present measurement.

A measurement process involves the actual physical operation of the specified equipment, following as closely as possible the procedures agreed on. It is subject to the many variations that can and do occur during the operation. The end result is an estimated best value that to be useful, must be accompanied by the uncertainty with reference to known performance parameters. Changes in any one or in a group of elements of the method constitute, in effect, a different particular method and a different process, which will in turn produce a different result and a different uncertainty.

Consequently, reliance is placed on procedures. This is why procedures are spelled out so meticulously in specifications, why there is exchange of procedural information, and why we are always anxious to learn about how the other fellow did it and what results he got. This utilization of established standard procedures and instruments is the clue to making reliable measurements, and it is one of the most important outcomes of all the considerations thus far.

Appendix D gives a typical summary sheet of specifications for triangulation, trilateration, traverse, and leveling. These are the standard National Geodetic Survey specifications for measurement, and they contain built-in checks and safeguards resulting from years of use. Familiar laboratory manuals of many varieties supply measurement procedures, testing methods, and so on, to establish for the user an assurance of precision (thus accuracy) if only a very few measurements or tests are to be made. Having followed established procedures, then, we are justifiably permitted to "adopt" as our own an established value of standard deviation for a single measurement.

Example 1. It has been established that by following "first-order" specifications, a level party can attain a difference of elevation within

± 0.017 ft$\sqrt{M}$, and by following "second-order" specifications, with ± 0.035 ft$\sqrt{M}$, where M is miles. Party A did 6 miles of first-order leveling from bench mark P to bench mark Q; party B, going from P to Q by another route, did 4 miles of second-order level work. Assuming that these formulas give limiting or maximum error, assign weights and adjust the elevation of bench mark Q, given the elevation of $P = 1471.612$ ft.

Party	Difference of elevation (ft)	Limiting error	Square of error	Adjustment factor	Adjustment	Adjusted values
A	49.018	$\pm 0.017\sqrt{6}$	0.0018	1.	+0.012*	49.030
B	49.064	$\pm 0.035\sqrt{4}$	0.0049	2.83	−0.034**	49.030
	0.046			$\Sigma = 3.83$	(0.0460)	

*0.012 = (1/3.83)0.046.
**0.034 = (2.83/3.83)0.046.

This example shows a lower-order measurement influencing one of higher order, something not often permitted in practice (for reasons extraneous to the present discussion). There is no reason to prohibit such a happening, provided the adjustment to each of the values lies within the range accepted as maximum or limiting error for each value. In this example these are:

Measurement	Accepted limiting error	Adjustment
First order	$\pm 0.017\sqrt{6} = \pm 0.042$	+0.0124
Second order	$\pm 0.035\sqrt{4} = \pm 0.070$	−0.0336

Adjustments of a magnitude that exceeds the range would be out of order: in such a case some blunder or undiscovered systematic error should be suspected in either or both sets of measurements.

The higher-order result (49.018) is known to be other than the true value, and it is a valid presumption that the lower-order result (49.064) can give some indication of where the true value lies with respect to the reported first-order value. The adjustment procedure is consequently that of Section 8.11.

The point to be made is that each of the measurement results has the validity it acquires by virtue of the established procedures used to obtain it. Therefore, each value can legitimately bestow its influence on the final result and validly contribute its weight in finding the adjusted mean.

The actual elevation of Q can be found from the known elevation of P by adding or subtracting the 49.030 ft. Notice that unless we can rely on established procedure for confidence in our precision, we are forced to perform great amounts of repetitious measurement.

It may be noted that the preceding technique is limited to situations of only two values or samples to be averaged. The following illustrates the general method, usable with any number (Section 8.11); it is shown with an additional party's input to the leveling work.

Party	Difference of elevation, X (ft)	Order of work	Miles	Limiting error, E	Weight ratio, $1/E^2$	Weight ratio, w	wX (omitting 49.0)
A	49.018	First	6	$\pm 0.017\sqrt{6}$	576.7	3.532	0.0636
B	49.064	Second	4	$\pm 0.035\sqrt{4}$	204.1	1.250	0.0800
C	49.037	Second	5	$\pm 0.035\sqrt{5}$	163.3	1.000	0.037
						$\Sigma = 5.782$	0.1806

Weighted mean:

$$\bar{X} = 49.0 + \frac{0.1806}{5.782} = 49.0 + 0.0313 = 49.0312$$

As is seen, this method allows for any number of values to be averaged, with due regard to the proper weight to be accorded to each.

Here is another example, illustrating the reliance that can be placed on a pattern of previously established results, the performance records of two measurement devices (or systems).

Example 2. In a measurement of a distance RS by two different electronic distance-measuring devices, each with its own well-established accuracy, a weighted mean can be ascertained from a set of measurements made by each. Instrument A: accuracy of 1 part in 300,000 of the length ± 2 in.; length reported as 3516.71 ft. Instruments B: accuracy of 5 parts in 1,000,000 of the length ± 0.04 ft; length reported as 3516.49 ft.

The establishment accuracies of each instrument can be used as the standard error of each value shown. Each value is in reality the mean of a set of four or six measurements. The individual errors are combined as follows:

$$\text{Instrument } A: \quad \sigma_m = \pm\sqrt{\left(\frac{3517\times 1}{3\times 105}\right)^2 + \left(\frac{2}{12}\right)^2}$$
$$= \pm\sqrt{0.000138 + 0.0278} = \pm\sqrt{0.0279}$$
$$= \pm 0.167 \text{ ft}$$

$$\text{Instrument } B: \quad \sigma_m = \pm\sqrt{\left(\frac{3517\times 5}{10^6}\right)^2 + (0.04)^2}$$
$$= \pm\sqrt{0.000309 + 0.0016} = \pm\sqrt{0.0019}$$
$$= \pm 0.044 \text{ ft}$$

Instru-ment	Value, X (less 3516.00)	σ_m	Weight ratio	Weight factor, w	wX
A	0.71	±0.167	$\left(\frac{1}{0.167}\right)^2 = 36$	1.	0.71
B	0.49	±0.044	$\left(\frac{1}{0.044}\right)^2 = 517$	14.4	7.06
				$\Sigma = 15.4$	7.77

Weighted mean:

$$\bar{X} = 3516.00 + \frac{7.77}{15.4} = 3516.50 \text{ ft}$$

We may reasonably accept this $\bar{X}$ value because both electronic instruments have been tested and used for many years according to the fixed pattern of procedure here employed, and because their results have been checked over and over against lengths that have been established firmly by other methods.

9.8 Setting Specifications for a Standard Procedure

To specify a measurement process involves ascertaining the limiting mean of the process, its variability due to random imperfections in the behavior of the system (i.e., its precision), the possible extent of systematic errors from known sources, or bias, and overall limits to the uncertainty of independent measurements.

Setting up a procedure for making measurements can be a tedious task, because a complete analysis must be made of the entire operation to determine the degree of leeway that can be accorded each step. Though we speak of analyzing from the desired precision back through to the actual procedure, the usual (and the simpler) way is to analyze forward, using the cut-and-try approach. A trial procedure is established, and the analysis of errors is carried through to establish the overall precision to be expected. If the precision is satisfactory, the procedure can be pronounced good; if too low, certain steps in the procedure must be tightened up; and if too high, certain steps may be relaxed. Out of such cut-and-try analysis will evolve a standard measuring procedure. Frequently, many attempts must be made before evolving a satisfactory measurement procedure that provides the desired precision of result.

Appendix C contains some extended analyses of procedures in measurement developed as indicated herein. By following these through, a clearer understanding of this treatment can be achieved.

9.9 Selecting the Number of Measurements To Be Made

To measure a value to some prescribed tolerance or accuracy, it becomes necessary to invert the operation with standard deviation and standard error so as to find the number of individual measurements to be made. This is perhaps best seen through some practical examples.

If we can measure an area by planimeter on a large-scale map to ± 0.006 in.^{2} (σ_s) in a set of eight trials, but the requirement is for ± 0.003 in.^{2}, it is probably obvious that we should repeat the measurement a considerable number of times.

Since for one set, $\sigma_s = \sqrt{\Sigma v^2/(8-1)} = \pm 0.006$ in.^{2}, it would seem that if we increase the denominator and not too greatly increase the numerator, we could expect to bring the σ_s down to the acceptable value of ± 0.003 in.^{2}. In the instance at hand, it may be that greater care in the use of the instrument, or greater proficiency, would do much to help. It is also possible to run around the area nonstop a number of times, say three times, with the expectation that the resulting area can be found by dividing by 3 and also cutting the σ_s to a third, thus

$$A = \frac{3A}{3} \pm \frac{\sigma_s}{3} = A \pm 0.002 \text{ in.}^2$$

But lacking this possibility, one may decide on making successive sets of eight trials each until a standard error (σ_m) can be obtained, that is, ± 0.003 in.^{2}. Let n be the number of sets in the formula:

$$\sigma_m = \frac{\sigma_s}{\sqrt{n}}$$

$$0.003 = \frac{\pm 0.006}{\sqrt{n}}$$

$$\therefore n = 4 \text{ sets}$$

or a total of 32 measurements in all. This is to be regarded with a bit of caution, nonetheless, for 8 is not a great number of values in a set. But one can become more comfortable through experience. Nothing can be better recommended than to take extra care and use a steady hand with the planimeter, for statistics will not compensate for sloppy workmanship.

Should a 95% assurance of the area result now be required, one would be obliged to consider that the ± 0.003 in.^{2} requested must be regarded as equal to $2\sigma_m$. Thus

$$2\sigma_m = \pm 0.003; \qquad \sigma_m = \pm 0.0015 \text{ in.}^2$$

and

$$\pm 0.0015 = \frac{\pm 0.006}{\sqrt{n}}; \qquad n = 16 \text{ sets (of 8 each)}$$

As suggested earlier, to avoid this inordinate number of measurements (128), efforts should be directed toward obtaining a more precise method of using the planimeter.

Ten

Two-Dimensional Errors

10.1 General

The previous work with normal distribution of random errors deals with only one-dimensional (linear) error theory. Measurements of lengths, elevations, angles, and such items have one-dimensional errors that can be handled by such theory. The principles of error theory can then be used advantageously to analyze the results and to fix the specifications for a survey or other measuring procedure.

To examine the accuracy of a position in a plane with respect to two axes, however, a further error analysis must be employed. The linear error component of two- and three-dimensional positions can be analyzed by applying the principles of normal linear error distribution. Two-dimensional error analysis is the subject of this chapter.

10.2 Definition

A two-dimensional error affecting a quantity is one defined by two random variables. For instance, the location of a point B set on the ground intended to be N5200.000 and E1400.000 is affected by two variables, direction and distance (or possibly by two distances only). In the (exaggerated) sketch of Fig. 10.1 it becomes apparent.

Since error exists (either plus or minus) in the measured length of AB and in the measured direction of AB, then the quadrilateral 1, 2, 3, 4 would

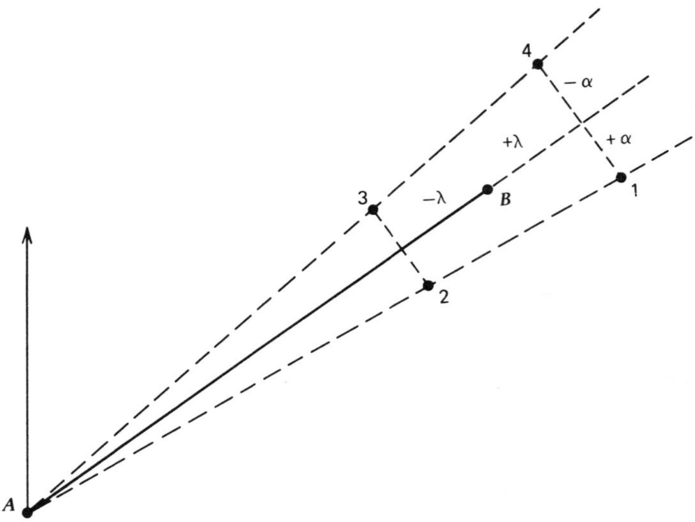

Fig 10.1 Position affected by length error and direction error: $+\lambda$ or $-\lambda$ is the error in distance; $+\alpha$ or $-\alpha$ is the error induced by the error in the angle.

seem to define the location of the possibly true position of B. Conversely, if we assume that B as plotted is the true or correct position of B, the quadrilateral may be regarded as the limit for the positioning of B when it is laid out.

10.3 The Probability Ellipse

If we consider that $\pm\lambda$ is the error in the distance and $\pm\alpha$ the error resulting from direction error, assuming that each is random and independent, then the probability (linear) density distribution of each error is:

$$p_\lambda = \frac{1}{\sigma_\lambda \sqrt{2h}} e^{-\lambda^2/2\sigma_\lambda^2}$$

$$p_\alpha = \frac{1}{\sigma_\alpha \sqrt{2h}} e^{-\alpha^2/2\sigma_\alpha^2}$$

The probability of each occurring simultaneously is $(p_\lambda)(p_\alpha)$, giving a two-dimensional probability density distribution:

$$(p_\lambda)(p_\alpha) = \frac{1}{(\sigma_\lambda)(\sigma_\alpha)2h} \exp\left[-\left(\frac{\lambda^2}{2\sigma_\lambda^2} + \frac{\alpha^2}{2\sigma_\alpha^2}\right)\right]$$

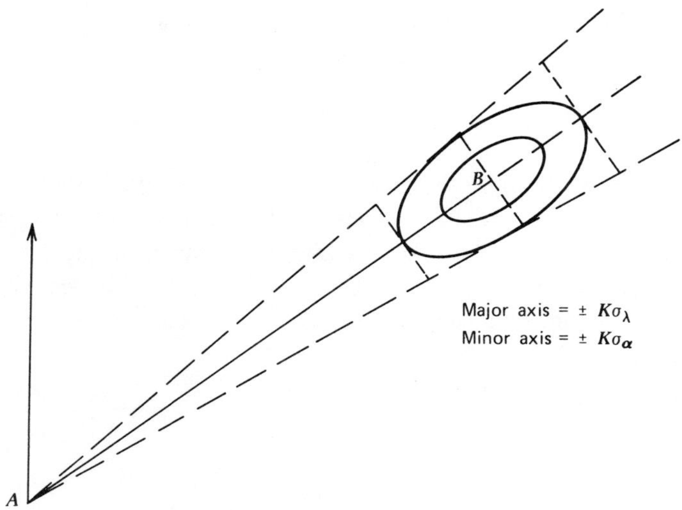

Fig 10.2 The probability ellipse.

This can be written as

$$(p_\lambda)(p_\alpha)\sigma_\lambda\sigma_\alpha 2h = \exp\left[-1/2\left(\frac{\lambda^2}{\sigma_\lambda^2} + \frac{\alpha^2}{\sigma_\alpha^2}\right)\right]$$

Therefore,

$$-2\ln(p_\lambda p_\alpha \sigma_\lambda \sigma_\alpha 2h) = \frac{\lambda^2}{\sigma_\lambda^2} + \frac{\alpha^2}{\sigma_\alpha^2}$$

Since for given values of p_λ and p_α, the left-hand member is a constant, then

$$K^2 = \frac{\lambda^2}{\sigma_\lambda^2} + \frac{\alpha^2}{\sigma_\alpha^2}$$

This shows that for values of $(p_\lambda)(p_\alpha)$ varying from zero to infinity, a family of equal probability density ellipses will occur with axes $k\sigma_\lambda$ and $k\sigma_\alpha$ (see Fig. 10.2).

10.4 The Probability Circle

But, if $\sigma_\lambda = \sigma_\alpha$, the equation can be seen to be that of a circle (really an equal-axis ellipse), since by substituting and rearranging terms, we obtain

$$-2\sigma_\lambda^2 \ln\left[p_\lambda p_\alpha \sigma_\lambda^2 (2h)\right] = \lambda^2 + \alpha^2$$

If we note that the left-hand member of the equation is a constant, say, $(k_1 r)^2$, the circular form becomes apparent:

$$(k_1 r)^2 = \lambda^2 + \alpha^2$$

Thus far we have been working with an error in length λ and an error in direction α. It now becomes apparent that if we utilize measurement methods to keep λ approximately equal to α, the resulting error ellipse becomes approximately an error circle. We may then expect our errors with respect to our coordinate axes to have exactly the same values, meaning that

$$\sigma_\lambda = \sigma_\alpha = \alpha_x = \sigma_y$$

(see Fig. 10.3).

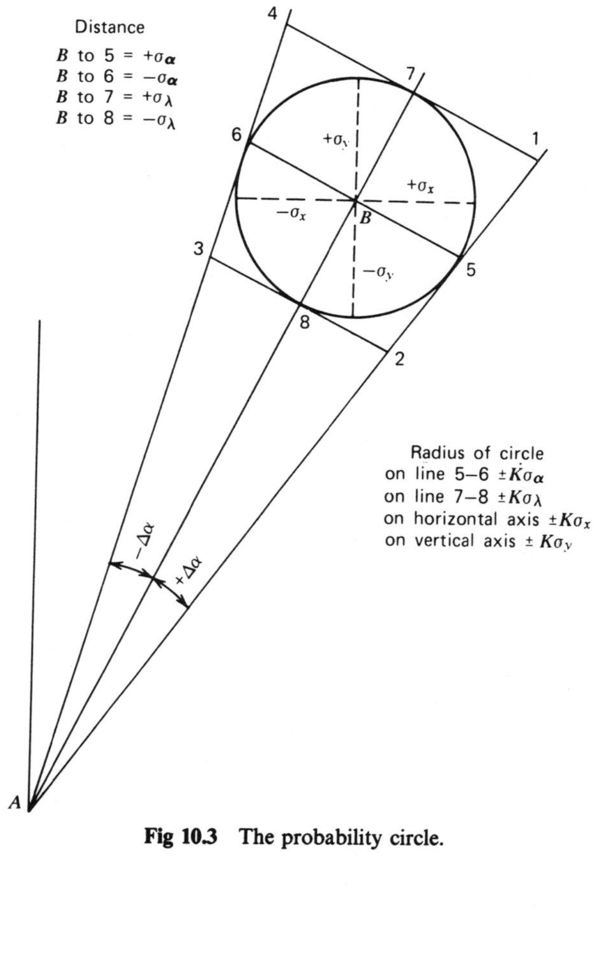

Fig 10.3 The probability circle.

We may then conclude that the error circle will have radius

$$k_2 = \sqrt{E_x^2 + E_y^2}$$

where E_x and E_y are any values of coordinate error. For convenience, let us call the errors in the X and Y coordinates x and y, respectively. We see then that the previous elliptical probability density equation can be properly written:

$$p_x p_y \sigma_x \sigma_y (2h) = \exp\left[-\frac{1}{2}\left(\frac{x^2}{\sigma_x^2} + \frac{y^2}{\sigma y^2} \right) \right]$$

By analogous reasoning, we arrive at the probability density circle:

$$(Kr)^2 = x^2 + y^2$$

10.5 Elliptical (Circular) Error Evaluation

When integrated, the probability density function gives the probability distribution function (the s curve). The probability of an ellipse is given by the distribution function:

$$P_{(x,y)} = 1 - e^{-K^2/2}$$

Solving this equation for various values of K yields these values for probability percentages in the accompanying table.

Probability, P (%)	K
39.3	1.000
50.0	1.177
63.2	1.414
90.0	2.146
99.0	3.035
99.8	3.500

The meaning is that when K is unity, for instance, 39.3% of all errors in a circular distribution will be within the limits of the circular standard deviation σ_c. This means, for example, that when $K=1.0000$, the axes of the ellipse are $1.0000\ \sigma_x$ and $1.0000\ \sigma_y$ (giving a circle of radius $\sigma_x=\sigma_y$) and that there is a 39.3% probability that the actual position errors in x and y will fall simultaneously within that circle. Increasing the diameter of the circle to $3.5000\ \sigma_c$ $(=3.5000\ \sigma_x=3.5000\ \sigma_y)$ will give a 99.8% probability that both the x and the y errors will fall within the circle.

Therefore, in the simple case of measuring to set out a point Q from point P, we must keep the length error consistent with the error resulting from direction.

Example 1 To set point Q with a maximum error of ± 0.05 ft, what shall be the limiting errors tolerated in distance and direction?

Strictly,

$$0.05\ \text{ft}=r=\sqrt{E_x^2+E_y^2}=\sqrt{E_L^2+E_a^2}$$

1. Since

$$E_L=E_a \qquad 0.05=\sqrt{2E_L^2}=E_L\sqrt{2}$$

and

$$E_L=\pm 0.035\ \text{ft}$$

This demands, therefore, a measurement precision ratio of $\pm 0.035/1000=1/28{,}000$. Reference to specifications (Appendix D) shows that this is a traverse of between first order and second order.

2. The angle (directional) error may not exceed

$$E_a=0.035\ \text{ft}=1000\sin\Delta\alpha \qquad \text{and} \qquad \Delta\alpha=\sin^{-1}0.000035$$

Since $\tan 01'=\sin 01'=0.00029$, and since small angles vary as their sines,

$$\frac{0.000035}{0.00029}=0.121'=\pm 07.3'' \text{ of arc}$$

Note that this angular error limitation assumes a single sighting 1000 ft long. Practical difficulties frequently render such a long sighting difficult or impracticable.

Example 2 In Example 1, if three instrument setups must be made to bring the final direction error to less than ±0.035, more careful work is needed. Let $E_a = \pm 0.035$ ft as before, but let each sighting now be 333 ft. This means effectively that the first sighting has an error in angle $\Delta\alpha$ that casts its effect 1000 ft, the second also a $\Delta\alpha$ that casts its effect 666 ft, and the third a $\Delta\alpha$ that casts its effect 333 ft. Effectively at point B, these displacement errors are, respectively, as shown in the accompanying table.

Setup	E	E^2
1	$\pm 1000 \sin \Delta\alpha$	$1{,}000{,}000 \sin^2 \Delta\alpha$
2	$\pm 666 \sin \Delta\alpha$	$440{,}000 \sin^2 \Delta\alpha$
3	$\pm 333 \sin \Delta\alpha$	$110{,}000 \sin^2 \Delta\alpha$
		$\Sigma = 1{,}550{,}000 \sin^2 \Delta\alpha$

Since

$$E_a = \pm\sqrt{E_1^2 + E_2^2 + E_3^2}$$

$$\pm 0.035 \text{ ft} = \sqrt{1{,}550{,}000 \sin^2 \Delta\alpha} = 1242 \sin \Delta\alpha$$

$$\sin \Delta\alpha = 0.000028$$

$$\Delta\alpha = \frac{0.000028}{0.00029} = \pm 0.0965' = \pm 05.8'' \text{ of arc}$$

This means that if we wish to keep the resulting position error within the known circular limits, we must use consistent distance and direction procedures. If in item 1 of Example 1 we wish to keep the position of point Q of our example 99.8% assuredly within a circle of 0.05 ft radius, we must acknowledge that the E_L and E_a (really the E_x and E_y also) are 3.5000 σ_x and 3.5000 σ_y. Consequently our measurements must be made (in the example) to $\pm 0.035/3.500 = \pm 0.010$ ft. Otherwise, as the example now stands, there will be only a 39.3% assurance.

10.6 Application to Position Accuracy

Now what does all this mean in the context of position accuracy? Must we always compute position error for individual points in question? Can the accuracy of these points be known? Or can we devise a means of stating position accuracy in a general way for all points in a system, be they set or located by traverse, triangulation, direct measurement, trilateration, or some other method.

To achieve consistency between distance precision and angular precision, we may simply note that this equality should obtain:

$$\frac{\sigma_\lambda}{\text{distance}} = \sigma_\alpha$$

The relative error in distance should equal the angular error in radians. The typical values in the accompanying table derive therefrom:

Precision in direction	Precision in distance
1°	1:57
10′	1:344
05′	1:690
01′	1:3,440
30″	1:6,870
20″	1:10,310
10″	1:20,620
01″	1:206,200

In the examples of Section 10.5, line *PQ* was used to set point *Q*, though no mention was made of the positional accuracy of point *P*. If we knew, for instance, that point *P* had a positional accuracy of ±0.05 ft in *X* and in *Y*, this error would have to be reckoned with in setting point *Q*. The $(E_x)_P$ and the $(E_x)_{PQ}$ must be summed thus to get $(E_x)_Q$:

$$\sqrt{(E_x)_P^2 + (E_x)_{PQ}^2} = (E_x)_Q$$

Of course this cannot be done for each line. Instead, a procedure is worked up for distance measurement and angle (direction) measurement to constitute a system that gives the desired accuracy. It is checked extensively to assure that the summation of all acccidental errors will not exceed a certain desired maximum value for 50% (or 90 or 99.8%) of the time. Specifications are spelled out in careful instructions, which then become the assurance of the resulting positional accuracy. Such a set of summary specifications is given in Appendix D for traversing and for triangulation. Working within the limits there prescribed can be expected to assure the desired positional accuracy.

10.7 Use of Control Systems

At this point it may be advisable to read the special section in Appendix E on the geoid, since it may be of assistance in understanding the present reasoning.

Essentially the question of positioning points on the surface of the earth (on the geoid, or on the ellipsoid adopted as its mathematical expression) is primarily a matter of using properly spaced and adequate control points. By some higher-order method such as astronomical observations, coupled with careful linking of observation stations by measurements between them, a control network can be established. The positional accuracy of these control stations can be computed from analysis of the errors of the astronomical and ground measurement methods. Then measurements and surveys of smaller areas can be linked to the larger well-controlled network, much as the skeleton of the body is the framework onto which and into which various bodily organs fit.

The use of such a control network thus precludes the errors that would tend to accumulate if the survey were to be extended piece by piece from one point outward without external check points. Whenever such a small survey is used and is extended from one starting point, a comparison of X and Y positions can be made at appropriate and available control stations of the network. Adjustments can be made to the better-fixed values of X and Y coordinates by common-sense methods, and the strength of the higher-order control network is thus utilized and extended throughout the minor system.

This notion of overall control is as essential to small operations as it is to nationwide or worldwide surveying and mapping. Carpenters and bricklayers set the corners of a building and work from these controlling points; a ship, an airplane, an automobile, or a large machine is similarly put together from controlling axes and points, not by the accretion method. The setting of this type of control for production operations is frequently done today by using the optical tooling procedure, which essentially utilizes the equipment and methods of surveying.

Bibliography

Moroney, M. J., *Facts from Figures*, 3rd ed., Penguin Books, New York, 1956.
A rollicking elucidation of statistics.

Crandall, Keith C. and Seabloom, Robert W., *Engineering Fundamentals in Measurements, Probability, Statistics, and Dimensions*, McGraw-Hill Book Company, New York, 1970.
Clear, elementary, and thorough; far from heartless.

Chatfield, Chirstopher, *Statistics for Technology*, Penguin Books, Inc., New York, 1970.
Thoroughly comprehensive, comprehensible, and practical.

U.S. Department of Transportation, FHA, BPR, "Quality Assurance of Portland Cement Concrete," a Research and Development Report (TD 2.109/a: H53), U.S. Government Printing Office, 1969.
A first indication of serious use of this newly developed statistical information theory in practice.

Baldridge, H. David, *Shark Attack*, Berkeley Medallion Book, Berkeley Publishing Corporation, New York, 1974.
A thoroughly absorbing book that rigorously illustrates statistical principles: it indicates what can't be done with data of varying degrees of reliability on some 1600 shark attacks, and draws some cautions conclusions.

Appendix A

Significant Figures in Measurement

A.1 Exact Versus Doubtful Figures

When a measurement is made, all digits in the result are *exact* if they are obtained by counting or by noting that a point lies between two markers, but digits are *doubtful* when they result from estimating. On a ruler graduated only in full inches, the digits 11 are exact; but if the height of a book can be measured as 11.2 in., the estimated 2 is doubtful or uncertain. However all three figures are significant. Significant figures include all exact digits and one doubtful digit. (It would be wrong and misleading in this case to call the book 11.21 in. or 11.213 in. high, since the smallest division on the ruler is 1 in. and the eye can at best estimate to only tenths. In this instance, too, we may say merely that it is 11.2 in., but certainly not 11.20 in. or 11.200 in. Why?)

A.2 Use of Significant Figures

In general, only figures or digits that are the result of actual measurement or calculation from an actual measurement are said to be significant (Values that result from counting are, of course, exact and are not here subject to discussion.) Rules and conventions, based on common sense, are detailed here.

A.3 Use of Zero

A-3-1

The zero is not significant when it serves merely to place the decimal.

1. 0.00275 m contains three significant figures; this would be clearer, perhaps, if written 2.75×10^{-3} m.
2. 54,000 miles has two significant figures (unless it is clearly intended that the value be exact). The use of the form 5.4×10^4 or 54×10^3 is not common, but it would be better usage in this respect.

A-3-2

When the zero is used otherwise, it is significant.

1. 5.7008 or 6.7760 or 6500.0 each has five significant figures.
2. 0.0077650 also has five significant figures, and it could well be written 77.650×10^{-4} for clarity.

A.4 Rules of Thumb for Significant Figures

A-4-1 Unless some precision index, such as sigma value or standard error, is affixed, as

$$a=1.755\pm0.003$$

the usual interpretation for the last (doubtful) figure is plus or minus one-half a unit in the last column. Thus a measured length of, say, 26.817 ft means that the range of uncertainty extends from 26.8165 to 26.8175 ft.

A-4-2 Though we expect and use only one doubtful figure in the final result, it is desirable to use two uncertain figures throughout the calculation and round off at the end.

A-4-3 Adding and subtracting: the sum or difference of several values is to be watched for doubtful figures. Note these obvious examples:

$$\begin{array}{r} 4.71 \\ 3.0 \\ 2.008 \\ \hline 9.7 \end{array} \qquad \begin{array}{r} 2.03 \\ 10.066 \\ 8.0412 \\ \hline 20.14 \end{array} \qquad \begin{array}{r} 178.612 \\ -2.1 \\ \hline 176.5 \end{array}$$

A-4-4 Multiplying or dividing: the result must not be credited with more significant digits than appear in the term with the smallest number of significant figures:

$$4.9178 \times 2.03 = 9.98 \qquad \text{(not } 9.983134\text{)}$$

$$(67.81 \times 10^3)^2 = 4{,}598 \times 10^6 \qquad \text{(not } 4598.1961 \times 10^6\text{)}$$

$$456.212 \div 2.17 = 210 \qquad \text{(not } 210.2359447\text{)}$$

Note, however, that 8 and 9 are nearly two-digit numbers, and occasionally an extra significant digit is thus warranted:

$$9.612 \times 3.00251 = 28.860 \qquad \text{(but not } 28.86012612\text{)}$$

A-4-5 The 10-in. slide rule is capable of just better than three digits.

A-4-6 In using logarithms, the number of decimal places required in the mantissa is fixed by the number of significant figures in the numbers being multiplied. Thus a six-place table is required for

$$326.712 \times 41.6123$$

A-4-7 Digital calculators sometimes convey a false notion of precision, and care must be taken to cut back and properly round off the final results. For example,

$$\tfrac{1}{3}(81.721 \times 9.3262) = 254.0487967 \qquad \text{or} \qquad 2.540487967 \times 10^2$$

on the hand calculator. Because the input values are measured quantities, of five significant digits each, the result can be properly shown to only five digits, thus 254.05, and not otherwise.

A.5 Rounding Off

When dropping excess digits, raise the last one to remain if the discarded quantity is greater than $\frac{1}{2}$, or leave it unchanged if the discarded quantity is less than $\frac{1}{2}$, thus:

3.476	becomes	3.48	or	3.5
3.512	becomes	3.51	or	3.5

If the quantity to be discarded is just 5, then round off the preceding digit to the nearest *even* value, thus:

4.875	becomes	4.88	(or 4.9)
4.885	becomes	4.88	(or 4.9)
4.8749	becomes	4.87	(or 4.9)
4.8851	becomes	4.89	(or 4.9)

A.6 Using Exact Values

Generally speaking, the preceding rules apply to quantities that result from measuring. If exact values are implied in a statement (e.g., "a 1500-ft radius circular curve"), the number of significant digits is not limited. Some instances are given here.

A-6-1 A field, measured by traversing, is computed to have an area of 9610.27 ft^2. It can be converted to acres by use of the *exact* conversion factor of 43,560 ft^2/acre

$$\frac{9610.27}{43560} = 0.220621 \text{ acre}$$

This could not, however, be rendered as 0.2206214417 acre, since it results from a measurement to only six digits.

A-6-2 If a circular curve of 850 ft radius is specified in a given case, this sets the degree of curve as

$$\frac{5729.57795}{850} = 6.7406799° \text{ (eight digits)}$$

or 6°44′26.45″ (which is equivalently seven or eight decimal digits). The validity of this comes from the ratio

$$\frac{D \text{ (degree of curve)}}{360°} = \frac{100 \text{ ft}}{2\pi R \text{ ft}}$$

where R is radius of curve, or

$$D = \frac{100 \times 360}{2(3.14159265)850}$$

the nine-digit value of π renders it possible to be sure of the eight-digit result. Note that being *exact*, 850 is equipped with an endless row of zeros, but the use of fewer digits for π would give erroneous results.

A-6-3 If the curve radius were specified as 85 ft (exactly), the degree of curve would be $D = 67.406799°$ (eight digits) or $D = 67°24'24.48''$ (also eight digits, approximately).

A-6-4 Trigonometric function tables are frequently used as though they furnish exact values, which they do not. For example, the sine or cosine of 45° may be listed as 0.70711 (to five digits) or 0.7071068 (to seven digits), but it is found on an electronic calculator to be 0.707106781 (to nine digits). Thus to obtain the Δx (departure) and Δy (latitude) of a survey line 2750.000 ft long, one will not get correct results unless at least seven (or better, eight) digits are used:

$$2750.000(0.70711) = 1944.553 \quad \text{(invalid)}$$
$$2750.000(0.7071068) = 1944.544 \quad \text{(valid)}$$
$$2750.000(0.70710681) = 1944.544 \quad \text{(valid)}$$

Note that the bearing of a line to tenths of a second has seven sexagesimal digits, approximately equivalent to seven decimal digits. Thus its sine and cosine functions would be properly obtained only from a seven-digit table or from an eight-digit or better calculator (but not from a six-digit calculator). Furthermore, some less expensive calculators tend to give a final digit that may be high or low when doing such trigonometric calculations.

Appendix B

Basic Concepts of Probability and the Normal Probability Curve

Linked to the notion of chance is the notion of probability (likelihood) of an occurrence such as the 50–50 chance of a tossed coin's coming up "heads." We can calculate such probabilities of events that are likely to occur, just as we can check the actuality of occurrences in controlled experimentation. Probability can be defined as the ratio of frequency of occurrence to the number of possible occurrences (i.e., number of successes to the number of trials).

The probability of an event lies between 0 and 1 (0% and 100%). The probability (p) of drawing a red card from a full deck is $26/52 = 0.50$ (or 50%). If an expert in sinking foul-line shots on the basketball court has sunk 891 of 1000, the probability of his putting in the next one is $p = 0.891$ (89.1%). With one die, the probability of rolling a number between 1 and 6 inclusive is 100% (or 1). The probability of rolling the number 5 is $\frac{1}{6}$ or 16.7%; the probability of its not being 5 is $100\% - 16.7\% = 83.3\%$.

The probability of *either* of two happenings (e.g., rolling a 5 *or* a 6) is the *sum* of their individual probabilities; in this case, $\frac{1}{6} + \frac{1}{6} = \frac{1}{3}$ or 33.3%.

The probability of two events occurring simultaneously (e.g., rolling a 5 *and* then a 6) is the *product* of the individual probabilities, in this case $(\frac{1}{6})(\frac{1}{6}) = \frac{1}{36}$ or 2.8%. Note that the probability of rolling a 5 and a 6 with two dice, however, is $(\frac{2}{6})(\frac{1}{6}) = \frac{1}{18} = 5.56\%$.

Table B.1 illustrates the probability concept, utilizing two dice rolled simultaneously. Plotting these probabilities in histogram form (Fig. B.1) helps illustrate visually the probability of rolling any given number (value).

Table B.1 Possible Combinations to be Rolled with Two Dice and Probability of Such Rolls

No.	All possible combinations	Probability	%
1	None	(0/6)(6/6) = 0/36	0.00
2	(1, 1)	(1/6)(1/6) = 1/36	2.78
3	(1,2), (2, 1)	(2/6)(1/6) = 2/36	5.56
4	(1,3), (2,2), (3, 1)	(3/6)(1/6) = 3/36	8.33
5	(1,4), (2,3), (3,2), (4, 1)	(4/6)(1/6) = 4/36	11.11
6	(1,5), (2,4), (3,3), (4,2), (5, 1)	(5/6)(1/6) = 5/36	13.89
7	(1,6), (2,5), (3,4), (4,3), (5,2), (6, 1)	(6/6)(1/6) = 6/36	16.67
8	(2,6), (3,5), (4,4), (5,3), (6,2)	(5/6)(1/6) = 5/36	13.89
9	(3,6), (4,5), (5,4), (6,3)	(4/6)(1/6) = 4/36	11.11
10	(4,6), (5,5), (6,4)	(3/6)(1/6) = 3/36	8.33
11	(5,6), (6,5)	(2/6)(1/6) = 2/36	5.56
12	(6,6)	(1/6)(1/6) = 1/36	2.78
13	None	(0/6)(6/6) = 0/36	0.00
		Σ = 36/36	100.00

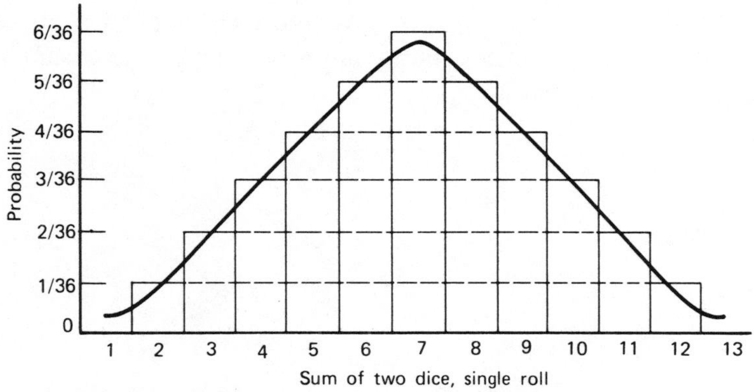

Fig B.1 Histogram showing probability of rolling any given number with a single roll of two dice (with the probability density curve plotted).

A smooth curve is superimposed to represent the limits of possibility, thus giving the *area* beneath the curve the probability value of 1.0 or 100%, as the column summations indicate. The probability of rolling a 9 or 10 or 11 is the *sum* of their respective probabilities (see columnar tally in table): 9/36 or 25%. This is the area beneath the curve, which corresponds to the numbers 9 to 11 inclusive, 9/36 or 25%. (Note that each box of Fig. B.1 represents an area of 1/36, and there is 36/36 = 1.0 or 100% certainty of rolling a number from 2 to 12 inclusive.)

If the number of dice thrown in a single roll were greater (say, about 10), the smooth curve would approach the bell-shaped "normal probability curve." Any segment of the area, then, beneath the probability curve represents the probability of an occurrence within the boundaries of the area.

DERIVATION OF NORMAL PROBABILITY CURVE

Let us suppose that we are recording a particular item (or parameter) of data, M, and that n readings of this parameter are taken. If each of these readings were taken with equal care, and a sufficient number of readings were taken, it can be said that the true value, as nearly as can be ascertained, or the most probable value of the parameter M, will be the arithmetic mean, that is,

$$\overline{M} = \frac{\Sigma M_n}{n}$$

By working with the equation above, we find that the number of readings multiplied by the arithmetic mean must be equal to the sum of the individual readings, that is,

$$n\overline{M} = \Sigma M_n$$

or

$$\overline{M} + \overline{M} + \overline{M} + \cdots + \overline{M} = M_1 + M_2 + \cdots + M_n$$

or

$$(\overline{M} - M_1) + (\overline{M} - M_2) + \cdots + (\overline{M} - M_n) = 0$$

but since $\overline{M}-M_n$ is the error of each individual reading and is denoted by X_n, we see that the sum of the errors is zero, or

$$x_1+x_2+x_3+\cdots+x_n=0$$

If we were now to consider two parameters, say, M and N, and make n observations of the functions of M and N, it would be possible to find the most probable value or arithmetic mean of M and N. The difference between particular observations and corresponding true values of the functions are the errors, each of which is a function of the parameters M and N. That is, errors are functions of the true value of a parameter. For example, there would be a larger absolute error in measuring a mile than in measuring a foot, generally speaking.

Now we can say that the probability of an error would be a function of the error. That is, if we denote probability by p, we have:

$$p_1=f(x_1)$$

$$p_2=f(x_2)$$

$$p_3=f(x_n)$$

When we say that the probability of an error is a function of the error, we mean that small errors are very probable and large errors are very improbable. A simple example will illustrate this. If we were measuring a distance of the order of 8 ft, an error of a foot would be highly improbable, whereas an error of an inch would be probable. Hence the probability of an error is a function of the magnitude of the error and this, in turn, is somewhat a function of the true value.

With a little thought it is easily ascertained that the probability of committing the given system of errors, that is, making all the possible errors simultaneously, would be

$$P=(p_1)(p_2)(p_3)\cdots(p_n)$$

Therefore,

$$p=f(x_1)f(x_2)f(x_3)\cdots f(x_n)$$

This means that since the absolute true value can never be ascertained by direct measurement, any measurement taken will have some error, even though it is small, hence we will always commit an error. Therefore, let us designate the probability of the whole system of errors as 1, or 100%.

If we plot the number of occurrences of each error against the particular value of the error, we get a curve something like Fig. B.2—assuming a usual occurrence pattern or distribution. Examination of the curve reveals that the probability of committing an error dx_1 is equal to the shaded area under the curve (i.e., the area $y_1 dx_1$). Thus

$$p_1 = y_1 dx_1$$

The probability density function integrated from $-\infty$ to $+\infty$ will yield 1 (or 100%). This is equivalent to stating that the area under the plotted curve contains 100% of all the errors.

It is obvious that if we let dx approach zero, the probability of the error will be proportional to the ordinate y. That is,

$$p_1 = ky_1$$

Let us now seek to design the equation of the normal probability curve in Fig. B.2. If we are interested in determining the most probable value of an error, we are actually interested in rendering the error a minimum. Mathematically, this is equivalent to equating the first derivative to zero. It was shown that

$$P = p_1 \cdot p_2 \cdot p_3 \cdots p_n \tag{1a}$$

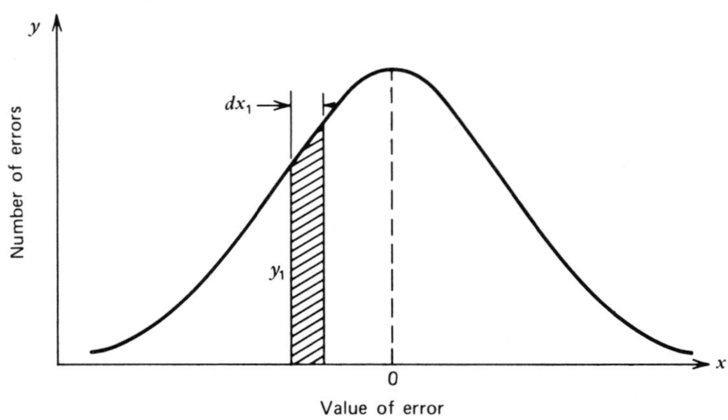

Fig B.2 Integration of the probability curve.

or

$$P=f(x_1)f(x_2)f(x_3)\cdots f(x_n) \tag{1b}$$

Now

$$f(x_1)=ydx=p$$

If we disregard the differential dx for the moment, we can write

$$y=f(x_1)f(x_2)f(x_3)\cdots f(x_n) \tag{1c}$$

or

$$\log y=\log f(x_1)+\log f(x_2)+\log f(x_3)+\cdots+\log f(x_n) \tag{2}$$

and taking the derivative of this equation with respect to M and equating to zero, we have

$$\frac{1}{y}\frac{dy}{dM}=\frac{1}{f(x_1)}\frac{df(x_1)}{dM}+\frac{1}{f(x_2)}\frac{df(x_2)}{dM}+\cdots \tag{3}$$

Now, there is a function $\phi(x)$ which when multiplied by $f(x)$ gives the derivative of $f(x)$. Therefore,

$$\frac{df(x)}{dx}=\phi(x)f(x) \tag{4a}$$

or

$$\phi(x)=\frac{df(x)}{f(x)dx} \tag{4b}$$

This can best be illustrated by an example. If we let

$$f(x)=2x^3$$

then

$$df(x)=6x^2dx$$

Since we said that

$$df(x)=\phi(x)f(x)dx$$

then if

$$\phi(x)=\frac{3}{x}$$

we have

$$df(x)=\frac{3}{x}2x^3dx$$

$$=6x^2dx$$

If we thus substitute equation 4b into equation 3, we get

$$\phi(x_1)\frac{dx_1}{dM}+\phi(x_2)\frac{dx_2}{dM}+\cdots=0$$

Now, if we say that the percentage error is the same no matter what the size of the measured quantity, we have

$$\frac{dx_1}{dM}=\frac{dx_2}{dM}=\frac{dx_3}{dM}=B$$

where B is a constant. Now, from before we know that

$$x_1+x_2+x_3+\cdots+x_n=0 \qquad (5)$$

and since

$$\frac{dx_1}{dM}=B$$

we then have

$$\phi(x_1)+\phi(x_2)+\phi(x_3)+\cdots+\phi(x_n)=0 \qquad (6)$$

It can be stated, from equations 5 and 6, that

$$\phi(x_1)+\phi(x_2)+\phi(x_3)+\cdots+\phi(x_n)=Cx_1+Cx_2+Cx_3+\cdots \qquad (7)$$

If we now substitute into equation 7 the value of $\phi(x)$ given in equation 4b, we have

$$\frac{df(x_1)}{f(x_1)dx_1}+\frac{df(x_2)}{f(x_2)dx_2}+\cdots=Cx_1+Cx_2+\cdots \qquad (8)$$

Since this must be true for any number of values, the corresponding values must be equal; therefore we have

$$\frac{df(x)}{f(x)dx} = Cx \tag{9}$$

If we integrate this, we have

$$\log f(x) = \tfrac{1}{2}Cx^2 + D$$

where D and C are constants; or by simplifying we have

$$f(x) = e^{(1/2)Cx^2 + D}$$

But now this is actually

$$f(x) = (e^{(1/2)Cx^2})(e^D)$$

But e^D is also a constant. Let us call this constant K, and we have

$$f(x) = Ke^{(1/2)Cx^2} = y \tag{10}$$

If we examine equation 10 we realize that C must be negative, since the probability must be decreasing as the value of x increases. That is, large errors are improbable. Let us let $C = -2h^2$ so that h and x are consistent. Then we have

$$y = Ke^{-h^2x^2} \tag{11}$$

It remains to evaluate the constant K. Since we desire the scale of our curve to be such that the ordinate y at any point will represent the probability of the size of an error given by x_1, and since we have stated that the sum of the probabilities of all possible errors must be one, we then must have

$$\int_{-\infty}^{+\infty} Ke^{-h^2x^2}dx = 1$$

Since the curve is symmetrical about $x = 0$, we then know that

$$\int_{0}^{+\infty} Ke^{-h^2x^2}dx = \frac{1}{2}$$

If we let

$$t = hx$$

$$dt = hdx$$

and

$$dx = \frac{dt}{h}$$

we have

$$\frac{1}{2} = \frac{K}{h}\int_0^{\infty} e^{-t^2}dt$$

It could be shown that the integral above

$$\int_0^{\infty} e^{-t^2}dt = \frac{\sqrt{\pi}}{2}$$

though it is beyond the scope of this text. Hence

$$\frac{1}{2} = \frac{K}{h}\frac{\sqrt{\pi}}{2} \qquad \text{or} \qquad K = \frac{h}{\sqrt{\pi}}$$

If we substitute this back into equation 11, we get

$$y = \frac{h}{\sqrt{\pi}} e^{-h^2x^2} \tag{12}$$

Recalling that we neglected the differential dx at the time, we can now say that

$$p = ydx = \frac{h}{\sqrt{\pi}} e^{-h^2x^2}dx \tag{13}$$

This is the equation of the normal probability density curve (one-dimensional).

If h evaluates as $1/\sqrt{2}\,\sigma_s$, then from the notion that the probability of an event is the area beneath the curve, the chance of a value of a normal distribution falling between $-\sigma_s$ and $+\sigma_s$ can be found (Fig. B.3).

$$p\Big|_{-\sigma_s}^{+\sigma_s} = \int_{-\sigma_s}^{+\sigma_s} y\cdot dx = \int_{-\sigma_s}^{+\sigma_s} \frac{1}{\sqrt{2}\,\sigma_s} e^{-\frac{x^2}{2\sigma_s^2}} = \frac{1}{\sqrt{2\pi}}\int_{-1}^{+1} e^{-(1/2)z^2}dz$$

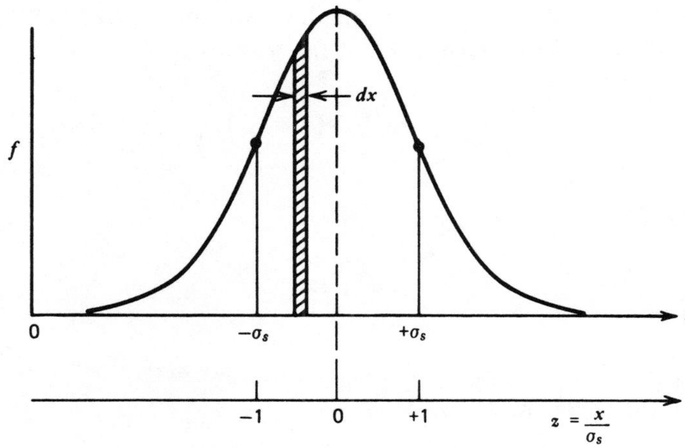

Fig B.3 Area evaluation for normal distribution curve.

Since

$$e^{(-1/2)z^2} = 1 - \frac{z^2}{2\cdot 1!} + \frac{z^4}{2^2\cdot 2!} - \frac{z^6}{2^3\cdot 3!} + \frac{z^8}{2^4\cdot 4!} - \frac{z^{10}}{2^5\cdot 5!} + \cdots$$

$$p = \frac{1}{\sqrt{2\pi}} \int_{-1}^{+1} \left(1 - \frac{z^2}{2} + \frac{z^4}{8} - \frac{z^6}{48} + \frac{z^8}{384} - \frac{z^{10}}{3840} + \cdots\right) dz$$

$$= \frac{1}{\sqrt{2\pi}} \left[z - \frac{z^3}{6} + \frac{z^5}{40} - \frac{z^7}{336} + \frac{z^9}{3456} - \frac{z^{11}}{42{,}240} + \cdots \right]_{-1}^{+1}$$

$$= \frac{1}{\sqrt{2\pi}} \left[2 - \frac{1}{3} + \frac{1}{20} - \frac{1}{168} + \frac{1}{1728} - \frac{1}{21{,}120} + \cdots \right]$$

$$= \frac{1}{\sqrt{2\pi}}(1.71125) = 0.6827 \quad \text{or} \quad 68.3\% \qquad \text{(Section 5.9)}$$

If the limits are $-2\sigma_s$ and $+2\sigma_s$, $p = 0.9546$ or 95.5% (Section 5.10).

PLOTTING THE NORMAL PROBABILITY CURVE

The normal probability curve can be plotted so that is is scale consistent with any histogran., to facilitate easy comparison, thus determining in any case that the sample distribution is normal. To do this, let us postulate that the curve can be demonstrated to be of the form

$$y=\frac{1}{\sqrt{2\pi}\,\sigma_s}\exp\left[-\frac{1}{2}(X-\bar{X})^2\sigma_x^{-2}\right]$$

This can be rendered in the form

$$y=\frac{KnI}{\sigma_s}$$

where I is the class interval of the histogram and the value of K can be taken from the following table.

x	K
$\bar{X}$ (mean)	0.39894
$\bar{X}\pm 0.5\ \sigma_s$	0.35206
$\bar{X}\pm 1.0\ \sigma_s$	0.24197
$\bar{X}\pm 1.5\ \sigma_s$	0.12953
$\bar{X}\pm 2.0\ \sigma_s$	0.05399
$\bar{X}\pm 2.5\ \sigma_s$	0.01753
$\bar{X}\pm 3.0\ \sigma_s$	0.00443
$\bar{X}\pm\infty$	0.00000

From the y-values obtained, a bell-shaped normal distribution curve consonant in scale with the sample data can be faired in. If it fits the histogram well, we can say that our data (our sample) are properly a representative sample of the whole population of measurements that could have been taken. This is the test. The judgment as to goodness of fit is somewhat subjective, being visual and without numerical standards readily at hand. But it is surprisingly simple and, with experience, one gains confidence in his judgment.

Example. The histogram of the 439 rod readings of Section 4.9 appears again in Fig. B.4. The actual distribution curve (dashed) is drawn by connecting the tops of the histogram bars. Then the normal distribution curve is calculated and plotted. Comparison of the two curves shows a

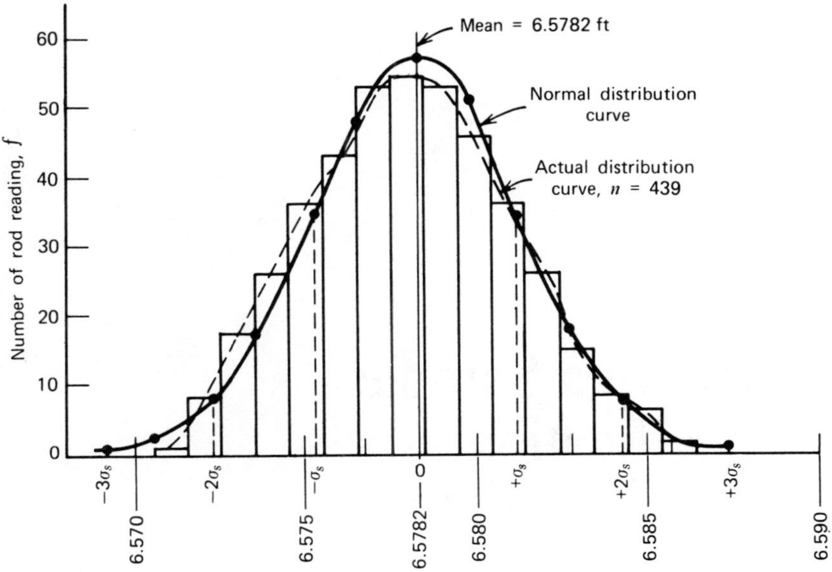

Fig B.4 Histogram of 439 rod readings with actual distribution curve and normal distribution curve plotted.

good fit, indicating that the sample is quite normally distributed. Here is the calculation of the ordinates of the normal curve:

$$y = \frac{KnI}{\sigma_s} = \frac{K(439)1}{0.00304} \qquad (\bar{X} = 6.5782)$$

x	y
$\bar{X}$	57.5
$\bar{X} \pm 0.5\ \sigma_s$	50.8
$\bar{x} \pm 1.0\ \sigma_s$	34.9
$\bar{X} \pm 1.5\ \sigma_s$	18.7
$\bar{X} \pm 2.0\ \sigma_s$	7.8
$\bar{X} \pm 2.5\ \sigma_s$	2.5
$\bar{X} \pm 3.0\ \sigma_s$	0.6
$\bar{X} \pm \infty$	0.0

Appendix C

Writing a Sample Specification for a Procedure

C.1 Taping Specifications

If we desire to design a procedure that will afford a precision of 1 : 10,000 when measuring with a steel tape, a good understanding of the equipment and its use, as well as of the systematic errors involved, is necessary.

Step 1. Let us assume (all errors being maximum) that:

(a) A good 100-ft steel tape (weight 0.0084 lb/ft) is available and can be compared with a standard to ±0.004 ft.

(*b*) A taping thermometer is available, reliable to ±1°F; and the air temperature recorded is not more than about ±3°F different from the tape temperature.

(*c*) The tape will be used with both ends supported, held horizontally, with a tension of 14 lb; it will be compared with the standard using this method of support at this tension.

(*d*) The tensiometer (spring balance) is available, reliable to ±1 lb.

(*e*) The end marking of tape lengths is done by suspending plumb bobs with an expertness that gives a maximum error of ± 0.007 ft.

(*f*) The tape ends are at the same elevation to within 0.5 ft.

(*g*) The tape has a cross section of 0.25×0.010 in.

Step 2. Analysis (see Table C.1).

(*a*) *Tape Comparison.* It is expected that any systematic error (difference in length between working tape and standard tape) is compensated for by a correction, and whatever error (E_1) remains is less than 0.004 ft. Though its sign is unknown, this error does nonetheless accumulate directly.

(*b*) *Temperature.* There may be an error of $\pm 3°$F in reported temperature at each tape length. This will be compensatory, not cumulative ($K = 0.00000645$ for steel).

$$E_t = KL(\Delta t)$$

Note. Some may validly argue that although we may miss the correct temperature value by being either too high or too low, thus making an error, the correct temperature will in reality be either consistently lower or consistently higher by reason of the consistency of the layers of air near the ground. Since this is true, they argue, the tape will have a consistent error in length, thus cumulative rather than compensatory. Obviously, on the basis of this assumption, the E_t should be regarded as cumulative—just as is the preceding E_l.

(*c*) *Pull.* The tensiometer may give an erroneous reading (or be read erroneously) by ± 1 lb for each tape length, and the sign is not known. This is compensatory, not cumulative ($E = 30 \times 10^6$ psi for steel).

$$E_p = \frac{PL}{AE}$$

The pull error in the table is a simple computation for a pull variation of 1 lb in the formula.

(*d*) *Sag.* The major (cumulative) sag error is already compensated in the tape comparison in paragraph *a*. If an error of ± 1 lb of tension is still possible, the sag will be plus or minus some computed value. This is compensatory, not cumulative.

$$E_s = \frac{W^2 L}{24P^2}$$

The sag correction must be computed for $P = 13$ lb and then for $P = 15$ lb. Half the difference of these two values gives an average sag error for ± 1 lb of P in that range.

(*e*) *Slope (gradient).* If one end of the tape is 0.5 ft higher or lower than the other when read, there will be the possibility of a distance error ranging from 0.000 to 0.125, always effectively shortening the tape. This will be cumulative, not compensatory (since its value is always on the same side of zero).

$$E_g = \frac{(\Delta h)^2}{2L}$$

(*f*) *Marking.* The ± 0.007-ft error is compensatory, not cumulative.

Step 3. Compute the estimated good procedure above (say, for 10,000 ft) and seek to ascertain the precision of the method.

Step 4. It becomes immediately apparent that this procedure is too meticulous for 1:10,000 precision, and certain specifications can be relaxed. The largest contributor of error is E_l. Note that if the tape comparison were made to ± 0.002 ft, the final E (all else being unchanged) would be ± 0.248 and the precision would show up as 1:40,000 for the procedure. For the present case, however, the procedure is uneconomically stringent for the precision desired, and the specifications may be amended, say, as follows:

(a) In tape comparison, find length to ± 0.006 ft.

(*b*) In tension, allow ± 2 lb.

(*c*) In temperature, expect to know within ± 5°F.

(*d*) In end marking, permit ± 0.01 ft.

This computation is shown in Table C.2.

Table C.1 Summary and Computation of Random Taping Errors: Computed for 10,000 ft or 100 tape lengths

Error, eh3Length,	± 0.004		Error for $\pm 0.004(100)$	0.160
E_L		Cumul.	$= \pm 0.4$	
Temperature,		Comp.	$\pm 0.0019\sqrt{100}$	0.00036
E_t	$\pm 0.00000645(100)(3) = \pm 0.0019$	(?)	$= \pm 0.019$	
Pull,			$\pm 0.0013\sqrt{100}$	0.000169
E_p	$\pm \dfrac{(1)(100)}{(0.25 \times 0.010)(30 \times 10^6)} = \pm 0.0013$	Comp.	$= 0.013$	
Sag,	$\pm \dfrac{1}{2}\left[\dfrac{(0.0084 \times 100)^2(100)}{24(13)^2} - \dfrac{(0.0084 \times 100)^2(100)}{24(15)^2}\right]$	Comp.	$\pm 0.0022\sqrt{100}$	0.00048
	$= \pm \dfrac{1}{2}(0.0174 - 0.0131) = \dfrac{0.0043}{2} = \pm 0.0022$		$= \pm 0.022$	
Slope,			$+0.00125(100)$	0.0156
E_g	$\dfrac{\pm (0.5)^2}{2(100)} = 0.00125$ (Cannot be minus in this context. Why?)	Cumul.	$= +0.125$	
Marking,	± 0.007	Comp	$\pm 0.007\sqrt{100}$	0.0049
E_m		.	$= \pm 0.070$	
E_m		.	$= \pm 0.070$	
Σ				$= 0.1816$

Total E:

$$\pm \sqrt{\Sigma E^2} = \pm \sqrt{0.1816} = \pm 0.425$$

Precision:

$$\pm \frac{0.425}{10{,}000} \text{-} \pm \frac{1}{23{,}000}$$

Table C.2 Step 4

Error, E	Value of error pertape length	Nature	Error for 10,000 ft, E	E^2
E_L	± 0.006	Cumul.	± 0.600	0.360
E_t	± 0.0032	Comp.	± 0.032	0.0010
E_p	± 0.0026	Comp.	± 0.026	0.0007
E_s	± 0.0028	Comp.	± 0.028	0.0008
E_g	± 0.00125	Cumul.	± 0.125	0.0156
E_m	± 0.01	Comp.	± 0.10	0.010

$\Sigma = 0.3881$

$$\text{Total } E: \pm\sqrt{\Sigma E^2} = \pm\sqrt{0.3881} = \pm 0.624$$

$$\text{Precision: } \frac{0.624}{10{,}000} = \frac{1}{16{,}000}*$$

*This is seen to be approaching the desired precision. Once a proper-sized error is anticipated per 10,000 ft (i.e., ± 1 ft), the specifications for the procedure should be carefully written.

Appendix D

Classification, Standards of Specifications and General Specifications of Geodetic Control Surveys National Ocean Survey, NOAA–February 1974

Table D.1 Standards for the Classification of Geodetic Control and Principal Recommended Uses.

Horizontal Control

Classification	*First-Order*	*Second-Order*		*Third-Order*	
		Class I	*Class II*	*Class I*	*Class II*
Relative accuracy between directly connected adjacent points (at least)	1 part in 100,000	1 part in 50,000	1 part in 20,000	1 part in 10,000	1 part in 5,000
Recommended uses	Primary National Network. Metropolitan Area Surveys. Scientific Studies	Area control which strengthens the National Network. Subsidiary metropolitan control.	Area control which contributes to, but is supplemental to, the National Network.	General control surveys referenced to the National Network. Local control surveys.	

Vertical Control

Classification	*First-Order*		*Second-Order*		*Third-Order*
	Class I	*Class II*	*Class I*	*Class II*	
Relative accuracy between directly connected points or benchmarks (standard error)	0.5 mm $\sqrt{K}$	0.7 mm $\sqrt{K}$	1.0 mm $\sqrt{K}$	1.3 mm $\sqrt{K}$	2.0 mm $\sqrt{K}$
			(K is the distance in kilometers between points.)		
	Basic framework of the National Network and metropolitan area control. Regional crustal movement studies. Extensive engineering projects. Support for subsidiary surveys.		Secondary framework of the National Network and metropolitan area control. Local crustal movement studies. Large engineering projects. Tidal boundary reference. Support for lower order surveys.	Densification within the National Network. Rapid subsidence studies. Local engineering projects. Topographic mapping.	Small-scale topographic mapping. Establishing gradients in mountainous areas. Small engineering projects. May or may not be adjusted to the National Network.

Table D.2 Classification, Standards of Accuracy, and General Specifications for Horizontal Control.

TRIANGULATION

Classification	*First-Order*	*Second-Order* Class I	*Second-Order* Class II	*Third-Order* Class I	*Third-Order* Class II
Recommended spacing of principal stations	Network stations seldom less than 15 km. Metropolitan surveys 3 km to 8 km and others as required.	Principal stations seldom less than 10 km. Other surveys 1 km to 3 km or as required.	Principal stations seldom less than 5 km or as required.	As required	As required
Strength of figure					
R_1 between bases					
Desirable limit	20	60	80	100	125
Maximum limit	25	80	120	130	175
Single figure					
Desirable limit					
R_1	5	10	15	25	25
R_2	10	30	70	80	120
Maximum limit					
R_1	10	25	25	40	50
R_2	15	60	100	120	170
Base measurement					
Standard error (1)	1 part in 1,000,000	1 part in 900,000	1 part in 800,000	1 part in 500,000	1 part in 250,000
Horizontal directions (2)					
Instrument	0".2	0."2	0".2 or 1".0	1".0	1".0
Number of positions	16	16	8 or 12	4	2
Rejection limit from mean	4"	4"	5" or 5"	5"	5"
Triangle closure					
Average not to exceed	1".0	1".2	2".0	3".0	5".0
Maximum seldom to exceed	3".0	3".0	5".0	5".0	10".0
Side checks					
In side equation test, average correction to direction not to exceed	0".3	0".4	0".6	0".8	2"
Astro azimuths (3)					
Spacing-figures	6-8	6-10	8-10	10-12	12-15
No. of Obser./night	16	16	16	8	4
No. of nights	2	2	1	1	1
Standard error	0".45	0".45	0".6	0".8	3".0
Vertical angle observations (4)					
Number of and spread between observations	3 D/R—10"	3 D/R—10"	2 D/R—10"	2 D/R—10"	2 D/R—20"

(Table D-2 continued)

Number of figures between known elevations	4-6	6-8	8-10	10-15	15-20
Closure in length (5) (also position when applicable) after angle and side conditions have been satisfied, should not exceed	1 part in 100,000	1 part in 50,000	1 part in 20,000	1 part ir 10,000	1 part in 5,000
			TRILATERATION		
Recommended spacing of principal stations	Network stations seldom less than 10 km. Other surveys seldom less than 3 km.	Principal stations seldom less than 10 km. Other surveys seldom less than 1 km.	Principal stations seldom less than 5 km For some surveys a spacing of 0.5 km between stations may be satisfactory.	Principal stations seldom less than 0.5 km.	Principal stations seldom less than 0.25.
Geometric configuration (6) Minimum angle contained within, not less than	25°	25°	20°	20°	15°
Length measurement Standard error (1)	1 part in 1,000,000	1 part in 750,000	1 part in 450,000	1 part in 250,000	1 part in 150,000
Vertical angle observations (4) Number of and spread between observations	3 D/R—10″	3 D/R—10″	2 D/R—10″	2 D/R—10″	2 D/R—20″
Number of figures between known elevations	4-6	6-8	8-10	10-15	15-20
Astro azimuths (3)					
Spacing-figures	6-8	6-10	8-10	10-12	12-15
No. of obs. night	16	16	16	8	4
No. of nights	2	2	1	1	1
Standard error	0″.45	0″.45	0″.6	0″.8	3″.0
Closure in position (5) after geometric conditions have been satisfied should not exceed	1 part in 100,000	1 part in 50,000	1 part in 20,000	1 part in 10,000	1 part in 5,000

TRAVERSE

Classification	*First-Order*	*Second-Order* Class I	*Second-Order* Class II	*Third-Order* Class I	*Third-Order* Class II
Recommended spacing of principal stations	Network stations 10-15 km Other surveys seldom less than 3 km.	Principal stations seldom less than 4 km except in metropolitan area surveys where the limitation is 0.3 km.	Principal stations seldom less than 2 km except in metropolitan area surveys where the limitation is 0.2 km.	Seldom less than 0.1 km in tertiary surveys in metropolitan area surveys. As required for other surveys.	
***Horizontal directions or angles* (2)**					
Instrument	0".2	0".2 or 1".0	0".2 or 1".0	1".0	1".0
Number of observations	16	8 or 12*	6 or 8*	4	2
Rejection limit from mean	4"	4" or 5"	4" or 5"	5"	5"
		* May be reduced to 8 and 4, respectively, in metropolitan areas.			
Length measurements					
Standard error (1)	1 part in 600,000	1 part in 300,000	1 part in 120,000	1 part in 60,000	1 part in 30,000
***Reciprocal vertical angle observations* (4)**					
Number of and spread between observations	3 D/R—10"	3 D/R—10"	2 D/R—10"	2 D/R—10"	2 D/R—20"
Number of stations between known elevations	4-6	6-8	8-10	10-15	15-20
Astro azimuths					
Number of courses between azimuth checks (7)	5-6	10-12	15-20	20-25	30-40
No. of obs night	16	16	12	8	4
No. of nights	2	2	1	1	1
Standard error	0".45	0".45	1".5	3".0	8".0
Azimuth closure at azimuth check point not to exceed (8)	1".0 per station or 2" $\sqrt{N}$	1".5 per station or 3" $\sqrt{N}$. Metropolitan area surveys seldom to exceed 2".0 per station or 3" $\sqrt{N}$	2".0 per station or 6" $\sqrt{N}$. Metropolitan area surveys seldom to exceed 4".0 per station or 8" $\sqrt{N}$	3".0 per station or 10" $\sqrt{N}$. Metropolitan area surveys seldom to exceed 6" per station or 15" $\sqrt{N}$	8" per station or 30" $\sqrt{N}$
***Position closure* (5)(8)** after azimuth adjustment	0.04m $\sqrt{K}$ or 1:100,000	0.08 $\sqrt{K}$ or 1:50,000	0.2m $\sqrt{K}$ or 1:20,000	0.4m $\sqrt{K}$ or 1:10,000	0.8m $\sqrt{K}$ or 1:5,000

(Table D-2 continued)

NOTE (1)

The standard error is to be estimated by

$$\sigma_m = \sqrt{\frac{\Sigma v^2}{n(n-1)}}$$

where σ_m is the standard error of the mean, v is a residual (that is, the difference between a measured length and the mean of all measured lengths of a line), and n is the number of measurements.

The term "standard error" used here is computed under the assumption that all errors are strictly random in nature. The true or actual error* is a quantity that cannot be obtained exactly. It is the difference between the true value and the measured value. By correcting each measurement for every known source of systematic error, however, one may approach the true error. It is mandatory for any practitioner using these tables to reduce to a minimum the effect of all systematic and constant errors so that real accuracy may be obtained.

"Manual of Geodetic Triangulation," Revised edition, 1959, for definition of "actual error."

NOTE (2)

The figure for "Instrument" describes the theodolite recommended in terms of the smallest reading of the horizontal circle. A position is one measure, with the telescope both direct and reversed, of the horizontal direction from the initial station to each of the other stations. See FGCC "Detailed Specifications" for number of observations and rejection limits when using transits.

NOTE (3)

The standard error for astronomic azimuths is computed with all observations considered equal in weight (with 75 percent of the total number of observations required on a single night) after application of a 5-second rejection limit from the mean for First- and Second-Order observations.

* See page 267 of Coast and Geodetic Survey Special Publication No. 247,

NOTE (4)

See FGCC "Detailed Specifications" on "Elevation of Horizontal Control Points" for further details. These elevations are intended to suffice for computations, adjustments, and broad mapping and control projects, not necessarily for vertical network elevations.

NOTE (5)

Unless the survey is in the form of a loop closing on itself, the position closures would depend largely on the constraints or established control in the adjustment. The extent of constraints and the actual relationship of the surveys can be obtained through either a review of the computations, or a minimally constrained adjustment of all work involved. The proportional accuracy or closure (i.e. 1/100,000) can be obtained by computing the difference between the computed value and the fixed value, and dividing this quantity by the length of the loop connecting the two points.

NOTE (6)

See FGCC "Detailed Specifications" on "Trilateration" for further details.

NOTE (7)

The number of azimuth courses for First-Order traverses are between Laplace azimuths. For other survey accuracies, the number of courses may be between Laplace azimuths and/or adjusted azimuths.

NOTE (8)

The expressions for closing errors in traverses are given in two forms. The expression containing the square root is designed for longer lines where higher proportional accuracy is required.

The formula that gives the smallest permissible closure should be used.

N is the number of stations for carrying azimuth.
K is the distance in kilometers.

Table D.3 Classification, Standards of Accuracy, and General Specifications for Vertical Control.

	First-Order	*Second-Order*		*Third-Order*
Classification	*Class I, Class II*	*Class I*	*Class II*	
Principal uses Minimum standards; higher accuracies may be used for special purposes	Basic framework of the National Network and of metropolitan area control Extensive engineering projects Regional crustal movement investigations Determining geopotential values	Secondary control of the National Network and of metropolitan area control Large engineering projects Local crustal movement and subsidence investigations Support for lower-order control	Control densification, usually adjusted to the National Net. Local engineering projects Topographic mapping Studies of rapid subsidence Support for local surveys	Miscellaneous local control; may not be adjusted to the National Network. Small engineering projects Small-scale topo. mapping Drainage studies and gradient establishment in mountainous areas
Recommended spacing of lines National Network	Net A; 100 to 300 km *Class I* Net B; 50 to 100 km *Class II*	Secondary Net; 20 to 50 km	Area Control; 10 to 25 km	As needed
Metropolitan control;	2 to 8 km	0.5 to 1 km	As needed	As needed
other purposes	As needed	As needed	As needed	As needed
Spacing of marks along lines	1 to 3 km	1 to 3 km	Not more than 3 km	Not more than 3 km
Gravity requirement;	0.20×10^{-3}gpu	....	....	....
Instrument standards (1)	Automatic or tilting levels with parallel plate micrometers; invar scale rods	Automatic or tilting levels with optical micrometers or three-wire levels; invar scale rods	Geodetic levels and invar scale rods	Geodetic levels and rods
Field procedures	Double-run; forward and backward, each section	Double-run; forward and backward, each section	Double- or single-run	Double- or single-run

(Table D-3 continued)

Section length	1 to 2 km	1 to 2 km	1 to 3 km for double-run	1 to 3 km for double-run
Maximum length of sight	50 m Class I; 60 m *Class II*	60 m	70 m	90 m
Field procedures (2)				
Max. difference in lengths				
Forward & backward sights				
per setup	2 m Class I; 5 m *Class II*	5 m	10 m	10 m
per section (cumulative)	4 m Class I; 10 m *Class II*	10 m	10 m	10 m
Max. length of line between connections	Net A: 300 km Net B: 100 km	50 km	50 km double-run 25 km single-run	25 km double-run 10 km single-run
Maximum closures (3)				
Section: fwd. and bkwd.	3 mm $\sqrt{K}$ Class I; 4 mm $\sqrt{K}$ Class II	6 mm $\sqrt{K}$	8 mm $\sqrt{K}$	12 mm $\sqrt{K}$
Loop or line	4 mm $\sqrt{K}$ Class I; 5 mm $\sqrt{K}$ Class II	6 mm $\sqrt{K}$	8 mm $\sqrt{K}$	12 mm $\sqrt{K}$

NOTE (1)

See text for discussion of instruments.

NOTE (2)

The maximum length of line between connections may be increased to 100 km for double run for Second-Order, Class II, and to 50 km for double run for Third-Order in those areas where the First-Order control has not been fully established.

NOTE (3)

Check between forward and backward runnings where **K is the distance in** kilometers.

Table D.4 National Geodetic Networks.

HORIZONTAL

Classification	Nationwide high precision traverses—Satellite Control	*First-Order*	*Second-Order* Class I	*Second-Order* Class II	*Third-Order* Class I	*Third-Order* Class II
Network component	Basic horizontal framework (control establishes the National Network)	Primary horizontal network (control develops the National Network)	Secondary horizontal control (control strengthens the National Network)	Supplemental horizontal control (Control contributes to the National Network)	Local horizontal control (control is referenced to the National Network)	
Nominal accuracy or precision between adjacent points	I part in 1,000,000	1 part in 100,000	1 part in 50,000	1 part in 20,000	1 part in 10,000	1 part in 5,000
Recommended density of control	Traverses and satellite stations at 900-1200 km. Stations at 15 km. to limit of technical and geometric restraints	Arcs not in excess of 100 km. Stations at 12-20 km. Urban control 3-8 km.	Stations at 10-13 km. Urban control 1-3 km.	As required	As required	

VERTICAL

Classification	*First-Order* Class I	*First-Order* Class II	*Second-Order* Class I	*Second-Order* Class II	*Third-Order*
Network component	Basic Vertical Network A (control establishes the National Network)	Basic Vertical Network B	Secondary Vertical Network (Control develops the National Network)	Supplemental Vertical Control (Control contributes to the National Network)	Local vertical control
Nominal accuracy *between points**	1.5 mm$\sqrt{K}$	2 mm $\sqrt{K}$	3 mm $\sqrt{K}$	4 mm $\sqrt{K}$	6 mm $\sqrt{K}$
Recommended density of *lines*	100-300 km	50-100 km.	25-50 km	10-25 km	As needed

* One-half of permissable closure

Appendix E

The Geoid

It is desirable to give here some basic concepts of the shape of the earth, the planet on which we live, because of our increasing need to know more about it. Though our species have inhabited the planet for thousands of years, we still have not ascertained all we need to know about its shape and size. The following discussion is a simplified explanation of some measurements made to find the shape of the earth.

A fluid mass rotating in space assumes the shape of a spheroid of revolution, the result of gravity force (self-attraction) and of centrifugal force (outward thrust of rotation). The resultant of the two forces is called potential. It is commonly, if incorrectly, referred to as gravity, simply because the centrifugal component is slight in comparison. If the mass is of uniform density throughout, the resultant shape is an ellipsoid of revolution and the potential is constant at all points on the surface. Thus we say that the ellipsoid is an equipotential surface.

In the case of earth, however, variations in topography and in specific gravity of materials which compose it cause the equipotential surface to deviate from an ellipsoid of revolution. The earth's equipotential surface, that surface which coincides most closely with mean sea level, is called the geoid. The geoid is always perpendicular to the direction of gravity. (See Fig. E.1.) The amount of variation or separation of the geoid surface from the mathematical ellipsoid surface is called the undulation of the geoid. Figure E.2 shows this.

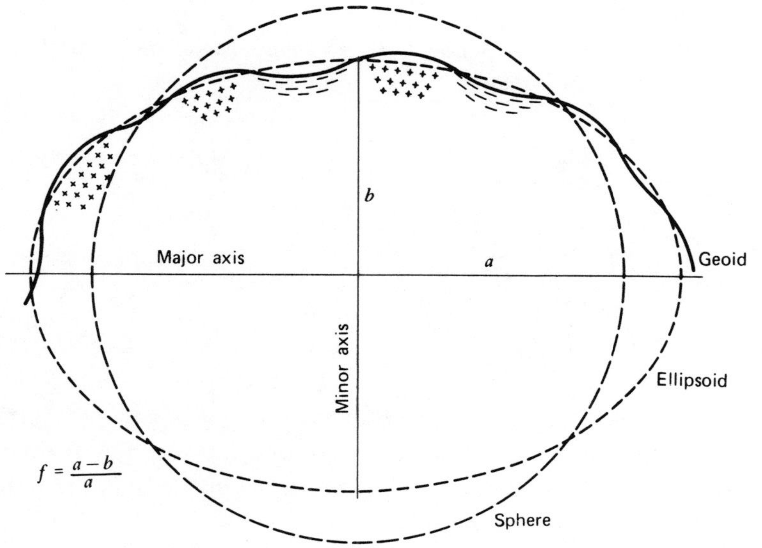

Fig E.1 Potential surfaces.

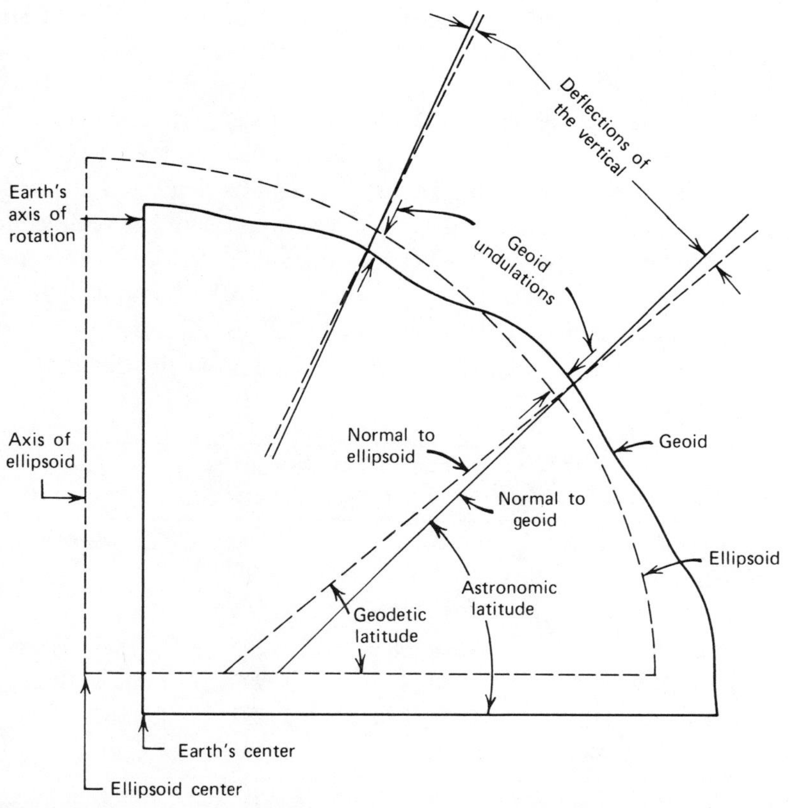

Fig E.2 Relationship of geoid to ellipsoid showing undulation of the geoid and deflection of the vertical.

Table E.1 Some Severe Cases of Deflections of the Vertical in the United States

Location	Deflection	
	Arc-seconds	Miles (approx.)
Columbus, Goergia	9.3	0.2^-
Point Arena, California	18.9	0.3^+
Salt Lake City, Utah	17.5	0.3^-
Forest Park, New York	7.2	0.1^+

An ellipsoid is generated by rotating an ellipse about its shorter axis. The size and shape of an ellipsoid is described by its radius at the equator a and its polar semi-diameter b. As indicated in Fig. E.1, the flattening is defined as $f=(a-b)/a$.

The geodetic position of a triangulation or traverse station is defined by the normal to the ellipsoid of reference. The astronomic position is defined by the plumb line at the station (the normal to the geoid). The difference of these two values is the deflection of the vertical as shown in Fig. E.2. Some sample values are given in Table E.1, together with the linear distances on the surface represented by the angular difference.

In principle, an earth ellipsoid is determined by taking astronomic observations along a meridian. These observations determine the angular difference in latitude. The length of the arc is measured by conventional geodetic triangulation or traverse. From the subtended angle and the arc length, the radius of curvature of the surface can be computed. In areas where the geoid is above the ellipsoid, the deduced radius will be too short, and conversely, where the geoid is below the ellipsoid the deduced radius

Table E.2 Ellipsoids in Current Use

Designation	Equatorial radius a	Flattening f	Where used
Airy (1830)	6,376,542	1/299	Great Britain
Everest (1830)	6,377,276	1/300	India
Bessel (1841)	6,377,397	1/299	Japan
Clarke (1866)	6,378,206	1/295	North America
Clarke (1880)	6,378,249	1/293	France, Africa
International (1924)	6,378,388	1/297	Europe
Krassowsky (1940)	6,378,245	1/298	Russia

will be too long. The reference ellipsoid is simply a least squares fit to a large number of such observations.

Several mathematical ellipsoids have till now been used in different areas of the world for geodetic computation. These are listed in Table E.2. Each was selected to fit the geoidal shape of a particular sector of the earth, though none manages to fit the entire earth very well. Figure E.3 shows, for instance, how the Clarke (1866) ellipsoid fits the North American continent well and the International (1924) fits Europe well. But it is evident that the two do not match where they meet, and one cannot perform distance and azimuth computations between the two unconnected geodetic systems.

The geoid, which any given ellipsoid tries to represent mathematically, is a surface along which the gravity potential (value of the gravity force) is everywhere equal, and to which the direction of gravity (vertical pull) is always perpendicular. This latter property is particularly significant because optical instruments containing leveling devices are commonly used to make geodetic measurements. The vertical axis of the instrument coincides with the direction of gravity (plumb line) and is therefore normal to the geoid, but not to the ellipsoid.

The geoid undulates above and below the most closely fitting ellipsoid in an erratic manner because of the uneven distribution of the earth's mass.

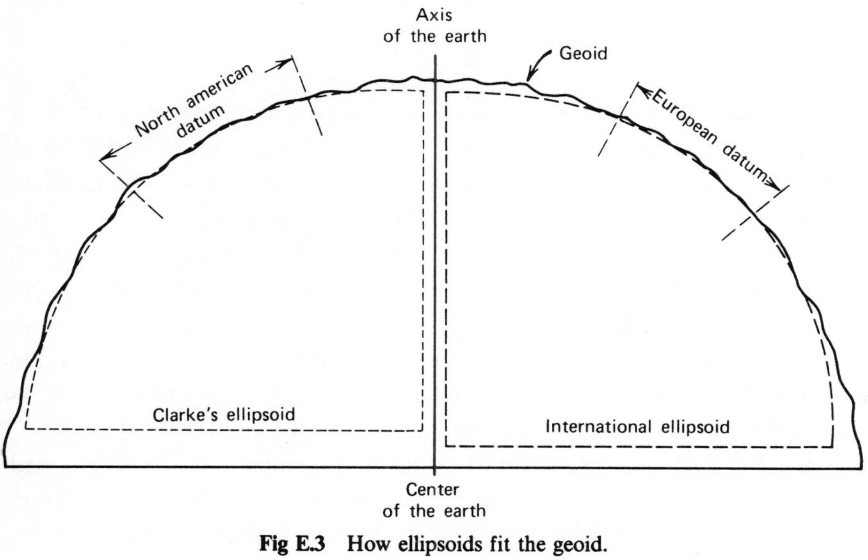

Fig E.3 How ellipsoids fit the geoid.

Concentrations of mass (e.g., mountains, or denser minerals) tend to "pull" the plumb line from the normal of the ellipsoid; similarly, deficiencies of mass (e.g., oceans, or especially ocean deeps) tend to "push" the plumb line. The problem then is to define the geoidal surface somehow, for it is with respect to this surface that we make our measurements.

It is not easy to ascertain the exact shape of the geoid, but great strides have been made. The easiest way is to measure the gravity or potential over the surface of the earth, that is, measure the pull of the earth on a particular mass at many locations. Extensive programs for gravity observation have been mounted by geodetic survey organizations of many nations.

Absolute gravity is found by a tedious and lengthy pendulum observation requiring many months at each station. However, finding the absolute value of gravity at a relative few base stations will suffice. Then relative gravity measurements are made at a great number of intermediate stations using a shorter and simpler procedure. This permits a great surrounding gravity field to be related to the base stations. Instrumentation breakthroughs have also been made in perfecting the gravimeter, which further eases the situation. It is now possible to measure relative gravity on shipboard or in moving aircraft, thus permitting rapid extension of worldwide gravity observations.

Gravity measurements reduced to sea level are compared with the theoretical gravity at that point on the ellipsoid of reference. Any difference is called the gravity anomaly for that point, and it converts directly into vertical distance of the geoid above or below the ellipsoid (undulation) as shown in Fig. E.4. Thus, point by point, the shape of the earth is determined.

The advent of satellites and ballistic missiles precipitated a need to establish a world geodetic system and a single ellipsoid which can be used worldwide. Fortunately, orbiting satellites carry their own geodetic blessing, for their orbits are determined directly by the gravity fields through which they pass. By careful tracking observations made on numerous spacecraft, gravity anomalies can be deduced from variations in the spacecraft orbits. Thus, the gravity field can be ascertained over great areas rather more rapidly than before. In addition, satellites afford the capability for determining geodetic coordinates of unknown positions from known stations by relatively unsophisticated tracking procedures.

If the deflection of the vertical cannot be determined, as on an isolated island, the position of the island as fixed by only astronomic observation can be greatly in error. This can occur if there exists local gravity anomaly,

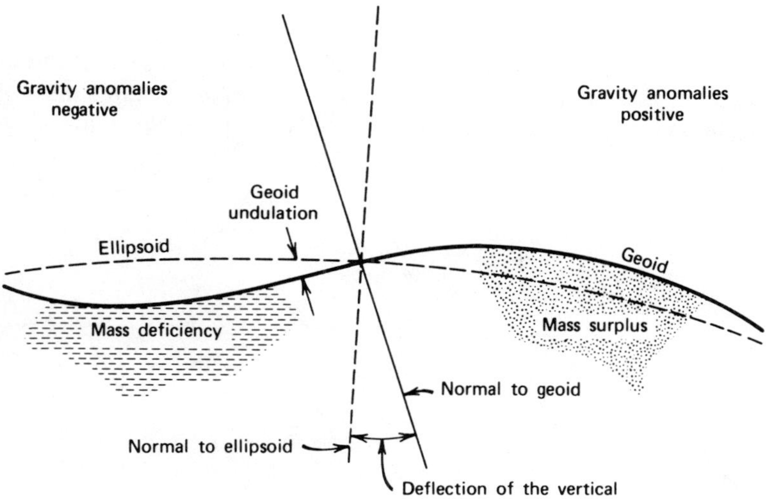

Fig E.4 Deflection of the vertical.

such as an extensive deep in the sea to one side and a huge land mass on the island itself. When certain islands are connected to continents by electronic distance-measuring methods (e.g., Shoran, Hiran), or by the new satellite observation methods, differences of as much as 5 or 6 miles are found between astronomic and geodetic position. Figure E.5 shows how an unknown station can be fixed in geodetic position by ranging on a satellite at the instant its position is fixed from three ground stations.

Today measurements of deflections of the vertical and of the value of gravity are being amassed to determine a new ellipsoid that will more

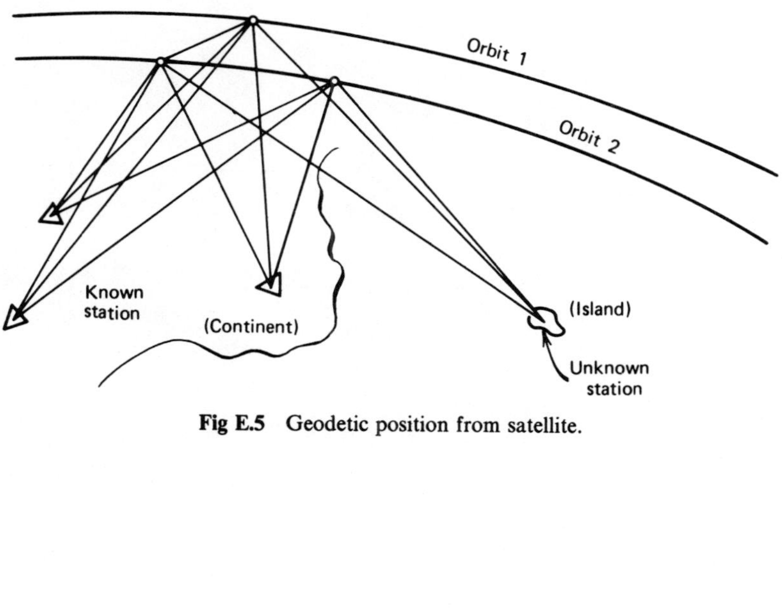

Fig E.5 Geodetic position from satellite.

closely fit the geoid worldwide. It will be known as the *world geodetic system ellipsoid* (Fig. E.6). Such a system may produce larger geoid undulations and deflections of the vertical within a continental land mass than previously. Once the system is initiated, however, all of the major geodetic networks of the world can be unified and the coordinates of points anywhere on earth will be compatible.

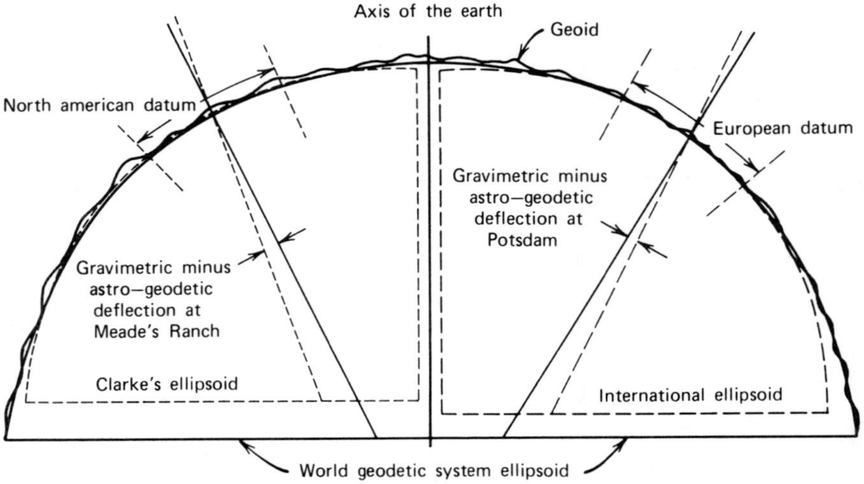

Fig E.6 World geodetic system ellipsoid as determined by deflections of the vertical computed from the earth's gravity.

Appendix F

The Frequency Distribution Curve

Though the *normal probability density* curve or the *normal distribution curve* has been adduced as a pattern for normalcy of data such as measurements of a quantity, this (bell-shaped) curve is both difficult to draw and difficult to analyze. Thus in practice it is not frequently used except to present the concepts of error distribution. Instead, we make judgments of sample distribution from the numerical data, from the histogram, or from simple rules of thumb based on experience.

It is possible to use also another type of curve, the S-curve or *frequency distribution curve*. Since this plots the *cumulative* frequencies, it sometimes is called the cumulative frequency distribution curve. Because a normal probability density curve can be rendered in this cumulative frequency form, such a curve is then called the *normal probability distribution curve*.

Confusion can exist, and care is necessary when the cumulative frequency notion is meant, since we talk glibly of the "normal distribution curve" (see Section 3.5) and of the "normal probability distribution curve" as we mean here. Of course "the normal probability curve" may designate either the distribution plot or the density plot. The distribution plot is the S-curve; the density plot is the bell curve.

The *frequency distribution curve* is essentially a plot of the cumulative number of values (or errors) against the same abscissa as before, the value of the quantity (or error). Using the cumulative technique for Example 4, Section 4.9 (the 439 observations) we have

Table F.1

Value, x (ft)	Number of occurrences, f	Residual (or variation), v (thousandths)	Cumulative number of occurrences	Percentage of occurrences	Adjusted* percentage of occurrences
6.571	1	−7	1	0.2	0.2
6.572	8	−6	9	2.0	2.0
6.573	18	−5	27	6.2	6.1
6.574	27	−4	54	12.3	12.3
6.575	36	−3	90	20.5	20.5
6.576	43	−2	133	30.3	30.2
6.577	53	−1	186	42.4	42.3
6.578	55	0	241	54.9	54.8
6.579	53	+1	294	67.0	66.8
6.580	46	+2	340	77.4	77.3
6.581	36	+3	376	85.6	85.5
6.582	26	+4	402	91.6	91.4
6.583	15	+5	417	95.0	94.8
6.584	13	+6	430	97.9	97.7
6.585	7	+7	437	99.5	99.3
6.586	2	+8	439	100.0	99.8
6.5782 (mean)	439 n				

*Explained subsequently.

The adjusted percentages of occurrences is explained thus. Since we wish to plot a finite number of an infinite set of data, insist the statisticians, we can never plot 100% of them. So, using the reasoning of Section 4.2, we here adjust the percentages for the plotting position of each error (or value). This is done simply by the formula

$$\frac{\text{cumulative number of occurrences}}{(n+1)} = \text{adjusted percentage}$$

and in Table F.1 the first few entries are:

$$\frac{1}{440} = 0.2\%, \qquad \frac{9}{440} = 2.0\%, \qquad \frac{27}{440} = 6.1\%$$

Plotting the cumulative distribution percentages against the values (or against the variations) gives the curve of Fig. F.1. This is the cumulative frequency curve or the frequency distribution curve, drawn in as a faired or smooth curve to fit the points. Superimposed are the values of $\pm\sigma_s$, which will normally fall on the central straight portion of the S-curve. The $\pm 2\sigma_s$ positions and the $\pm 3\sigma_s$ positions do not. These might be checked, incidentally, against the percentages of Sections 5.9 through 5.11 to see if they contain the correct percentages of occurrences (see also Table 5.3). They would, of course, check properly in a theoretically correct normal distribution of an infinite number, and fairly well in any good distribution of errors (or values) in our usual samples.

The cumulative frequency curve (the S-curve of Fig. F.1) is also known as the "less than/more than" curve. For instance, 90% of all the readings are less than (smaller than) 6.5818 ft. If the numbers represented tight-fit inserts to fit snugly in a bore of 6.580, it can be seen that only 77% would fit without being cooled for shrink fitting. Highway departments use the curve to establish good limiting speeds for segments of highways, selecting as a limit the speed that 85% of the drivers travel. This means that 15% who travel above that speed are not good representatives of safe driving

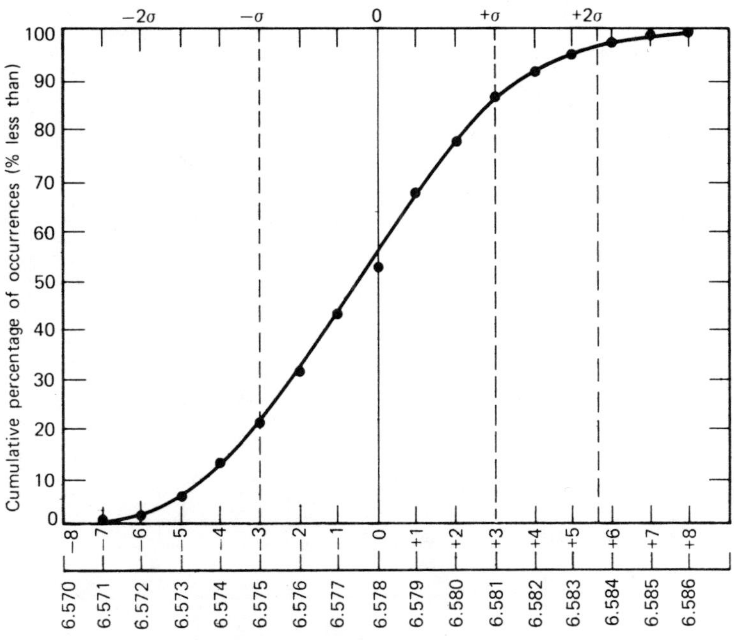

Fig F.1 Cumulative frequency distribution curve for the 439 rod readings.

practice and should be curbed. The 85% mark in any normal distribution represents the point where the S-curve starts to turn off to the right as it rises.

Still simpler methods of representing a frequency distribution have been developed, and one involves the use of *arithmetic probability paper*. The abscissa is the same as before, but the ordinate (%) scale is expanded at both top and bottom to straighten out the loops of the "S" of the preceding frequency distribution curve. Such arithmetic probability paper is available in several slightly different forms. Various attempts have been made by different people to design a form that portrays a normal distribution as a straight line, sometimes for specialized purposes. The form in Fig. F.2 depicts our 439 rod readings.

The result of plotting a normal distribution of errors (or values) on this arithmetic paper is a straight line. Conversely, plotting one's sample values on such paper and achieving a straight line indicates that the distribution is normal. In our case, some 15 points have been fitted on the paper and connected by a dashed line—which is a fairly straight line. Then a straight line is drawn, the line of *best fit*, which is the theoretically best representation of the sample. It is drawn through only two (or three) known values: (1) the mean value, (2) the $+\sigma_s$ value (and the $-\sigma_s$ value as a check).

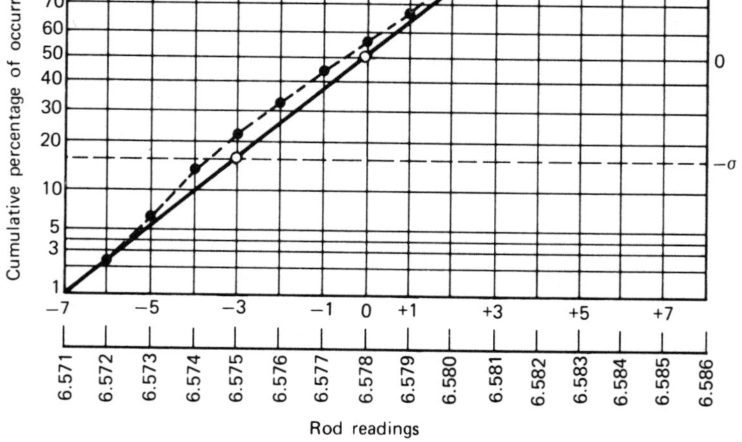

Fig F.2 Set of 439 rod readings plotted on arithmetic probability paper.

Examination of the two curves (lines) shows good matching or correlation, which indicates that the 439 rod readings are pretty well evaluated as a normal distribution.

It is, in fact, this arithmetic probability paper plot of the frequency distribution curve that is accepted almost universally as a test of the normalcy of the data gathered in measurements. And it is the curve from which one makes any studies and draws conclusions. Thus by using the standard deviation and the mean for the plot, we can obtain a good visualization of the distribution of values, of scatter, and of precision. One need not plot more values generally.

An additional notion to be established here is that we can use the arithmetic probability paper just as simply and assuredly with the mean and the standard error (σ_m), for the standard error is the sigma value of the mean. In this context we are then talking of the whole population as opposed to a particular sample. In fact, most often we are dealing with data for a small batch, and we should be plotting the σ_m on the paper. Only when from the context it is clear that we already have the standard error (and the label given to it is "standard deviation" by those furnishing the data) should we avoid further seeking to render it into standard error form by dividing by $\sqrt{n}$.

A further example is given here, using results of daily tests made on the effluent from an industrial waste treatment plant during the month of April. At issue is the efficiency of the removal process.

Example. The percentage of biological oxygen demand (BOD) for each day of the month was found at one of West Virginia Pulp and Paper Company's industrial waste treatment plants. From the data given (percentage of removal), determine the mean for the month, the standard deviation, and so on, and plot the data on arithmetic probability paper.

April	1	80.7%	April	6	80.2%	April	11	81.6%
	2	87.6%		7	72.6%		12	79.3%
	3	87.0%		8	78.4%		13	79.3%
	4	82.6%		9	72.1%		14	80.1%
	5	83.0%		10	79.6%		15	72.6%
April	16	74.8%	April	21	78.4%	April	26	71.4%
	17	73.6%		22	80.4%		27	73.3%
	18	78.8%		23	78.2%		28	75.9%
	19	75.3%		24	81.5%		29	79.8%
	20	77.7%		25	79.1%		30	79.9%

Table F.2.

Value, X	Cumulative number	Variation, v	fv^2	Adjusted percentage occurrence
71.4	1	−7.2	51.84	3.2
72.3	2	−6.3	39.69	6.5
72.6	3	−6.0	36.00	9.7
73.3	4	−5.3	28.09	12.9
73.6	5	−5.0	25.00	16.1
74.8	6	−3.8	14.44	19.4
75.3	7	−3.3	10.89	22.6
75.9	8	−2.7	7.29	25.8
76.2	9	−2.4	5.76	29.0
77.7	10	−0.9	0.81	32.3
78.2	11	−0.4	0.16	35.5
78.4	12	−0.2	0.04	38.7
78.4	13	−0.2	0.04	41.9
78.8	14	+0.2	0.04	45.2
79.1	15	+0.5	0.25	48.4
79.3	16	+0.7	0.49	51.6
79.3	17	+0.7	0.49	54.8
79.6	18	+1.0	1.00	58.1
79.8	19	+1.2	1.44	61.3
79.9	20	+1.3	1.69	64.5
80.1	21	+1.5	2.25	67.7
80.2	22	+1.6	2.56	71.0
80.4	23	+1.8	3.24	74.2
80.7	24	+2.1	4.41	77.4
81.5	25	+2.9	8.41	80.6
81.6	26	+3.0	9.00	83.9
82.6	27	+4.0	16.00	87.1
83.0	28	+4.4	19.36	90.3
87.0	29	+8.4	70.56	93.5
87.6	30	+9.0	81.00	96.8
$\Sigma X = 2{,}358.4$		$\Sigma_v^2 = 442.24$		

Mean

$$\bar{X} = \frac{2{,}358.4}{30} = 78.6$$

Standard deviation:

$$\sigma_s = \sqrt{\frac{\Sigma v^2}{n-1}} = \sqrt{\frac{442.24}{29}} = \sqrt{15.25} = \pm 3.91$$

Table F.2 is the frequency distribution table. Because very few values occur more than once, this table differs slightly from that of Section 4.8, where there were many values repeated.

The curve is now plotted on arithmetic probability paper simply by plotting the mean (78.6) and the $+\sigma$ value ($78.6+3.9=82.5$); the $-\sigma$ value is also plotted as a check ($78.6-3.9=74.7$).

By also plotting the individual values, it will become apparent that the data of this problem are distributed in a fairly normal pattern. Were this not so, it would become evident after a little experience with such plotting.

The frequency distribution tabulation sheet and the arithmetic probability paper on the pages that follow may be photocopied for use without the publisher's permission.

Frequency Distribution Tabulation Sheet

Values, X	Tally space	f	fX	v	fv	fv^2
Total [Σ]						

From B. Austin Barry, ***Errors in Practical Measurement in Science, Engineering, and Technology***, M.D. Morris (Ed.), John Wiley & Sons, New York, 1978

Arithmetic probability paper

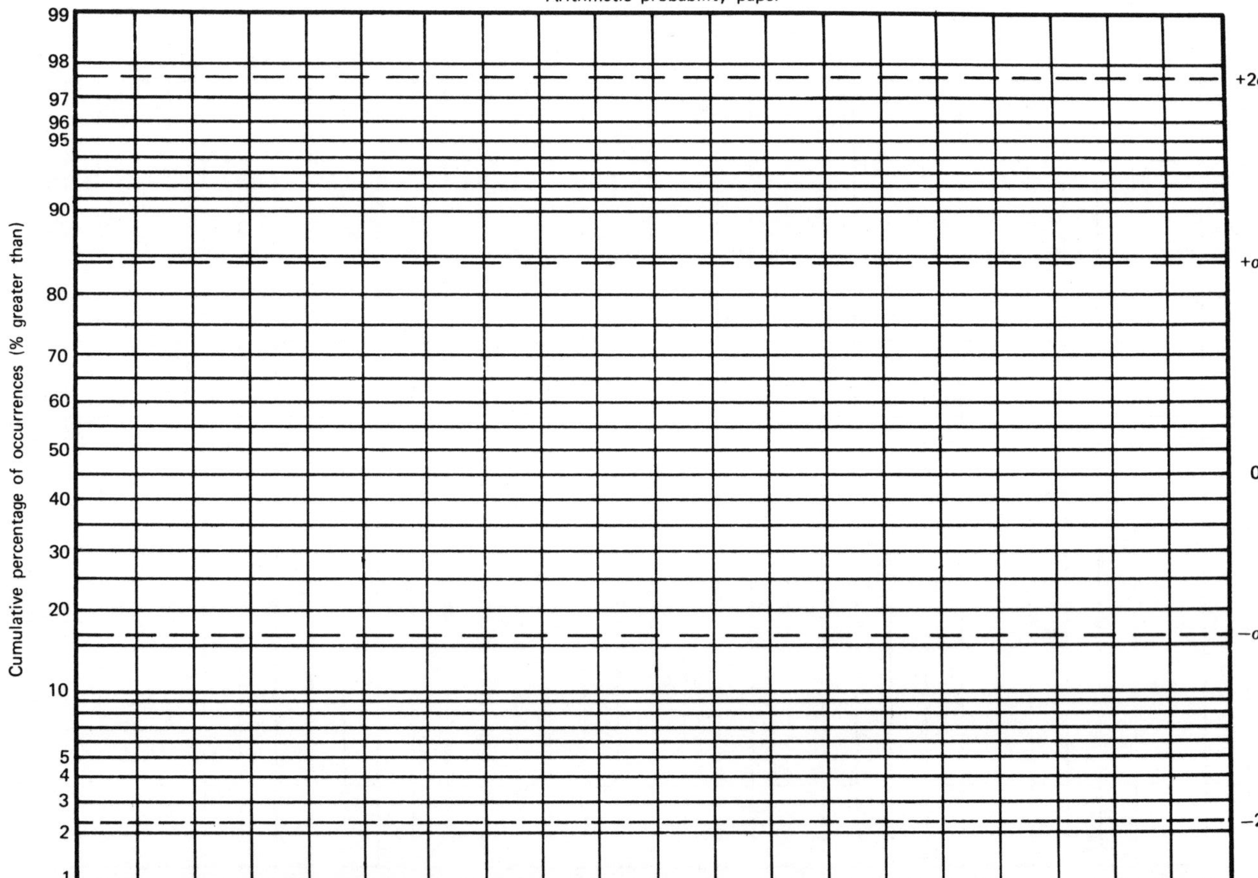

From B. Austin Barry, ***Errors in Practical Measurement in Science, Engineering, and Technology***, M.D. Morris (Ed.), John Wiley & Sons, New York, 1978

Appendix G

Computer and Calculator Solution of Problems

Since digital computers have come into our lives, offering new facility for solution of problems and handling of data, some notions of their programming and operation are given here. A simple standard deviation problem is included, with a solution by a Fortran program on a computer (Fig. G.1), followed by an explanation of the working of the hand calculator to accomplish the same thing.

Example Following are three groups of data, 50 items each. Find the standard deviation for each group, the mean for each group, and the standard error of the mean.

Group 1

56.9	57.1	54.4	58.0	57.1
58.1	57.7	55.2	58.1	57.9
54.5	58.8	56.8	57.6	57.1
54.0	56.5	55.6	57.1	58.0
57.1	57.2	57.1	57.9	58.9

57.8	58.0	54.4	58.0	59.8
57.3	58.3	54.9	57.3	58.2
57.9	58.5	54.5	56.9	57.9
58.2	58.1	54.3	57.9	57.5
56.8	57.0	54.6	57.5	57.2

Group 2

51.8	54.5	53.5	52.1	51.9
49.1	54.2	52.1	51.3	52.6
51.9	54.8	53.6	51.1	52.4
51.1	54.8	55.7	53.9	51.3
51.5	52.0	54.5	51.0	52.4
52.8	53.0	54.6	51.0	53.0
49.9	53.3	54.9	52.1	52.7
51.3	52.0	54.3	53.8	53.2
53.0	53.2	53.9	53.5	52.0
52.5	54.8	53.3	51.5	51.6

Group 3

56.9	57.1	54.4	58.0	57.1
58.1	57.9	55.2	58.1	57.9
54.5	58.8	56.8	57.9	57.1
54.0	56.5	55.6	57.1	58.0
57.1	57.2	57.1	57.9	58.9
57.8	58.0	54.4	58.0	58.9
57.3	58.3	54.9	57.3	59.8
57.9	58.5	54.5	56.9	58.2
58.1	58.1	54.3	57.9	57.9
56.8	57.0	54.6	57.8	57.5

Since most computer installations have a standard deviation subroutine in their program files, such a program need not be written any longer, merely summoned when desired.

```
FORTRAN IV      V01C-03H    TUE 07-MAR-78 14:45:01                    PAGE 001

        C       >>>>>>>>>> Manhattan College Student Batch System  V3.0 <<<<<<<<<<
        C       LOADED: 07-Mar-78       02:44 PM        JOB ID: G53094
        C       $FOR (100,1) $SYSTEM
        C
        C       PROGRAMMER: UNKNOWN USER        ACCOUNT: (100,1)
        C
0001            CALL ASSIGN (1,'G53094.DAT')        ! READ
0002            CALL ASSIGN (6,'G53094.OUT/C')      ! PRINT
        C
        C       ***************************************************************
        C       MANHATTAN COLLEGE A K
0003            DIMENSION A(150)
0004    1       READ 1001, N
0005            IF (N .EQ. 999) GO TO 4
0007            READ 1000, (A(I), I = 1, N)
0008            SUMA = 0.0
0009            DO 2 I = 1, N
0010    2       SUMA = SUMA + A(I)
0011            SN = N
0012            SMEAN = SUMA / SN
0013            SUMB = 0.0
0014            DO 3 I = 1, N
0015            SVAR = A(I) - SMEAN
0016    3       SUMB = SUMB + SVAR * SVAR
0017            SIGMAS = SQRT(SUMB / SN)
0018            SIGMAM = SIGMAS / SQRT(SN - 1)
0019            PRINT 1002, SMEAN, SIGMAS, SIGMAM
0020            GO TO 1
0021    1000    FORMAT (5F4.1)
0022    1001    FORMAT (1I3)
0023    1002    FORMAT (6H MEAN=, F9.5, 5X, 9HSTAN DEV=,
               1F9.5, 5X, 11HSTAND ERR =, F9.5)
0024    4       END

Begin Program Execution

MEAN= 56.98800      STAN DEV=  1.28602      STAND ERR =  0.26251
```

Fig G.1

BY ELECTRONIC CALCULATOR

Some electronic pocket or desk calculators are wired with a simple program to find the standard deviation (σ_s) of a sample. The values are entered successively on the keyboard and for each the "Σ" (or "$\Sigma+$") key is depressed. When all have been entered, another key (e.g., "$\overline{X}$") is tapped, to carry the program to completion.

The computation is based upon the formula using an assumed mean

$$\sigma_s = \sqrt{\frac{n}{n-1}}\ \sqrt{\frac{\Sigma v'^2}{n} - \left(\frac{\Sigma v'}{n}\right)^2}$$

but the mean assumed is *zero* (from which v' is found in each case). This effectively reduces the formula to

$$\sigma_s = \sqrt{\frac{n}{n-1}}\sqrt{\frac{\Sigma X^2}{n} - \left(\frac{\Sigma X}{n}\right)^2}$$

This gives large numbers to handle (something, of course, to be avoided in hand work), but it is not a great problem for the calculator.

The hard-wired program in the hand calculator is designed simply so that as one inputs each successive value (X) of the sample, three things happen:

1. The value X is added to register ①,
2. The value X^2 is added to register ②,
3. The value 1 is added to register ③.

This will allow the calculation of σ_s after the entering of any value, by programming the contents of registers ①, ②, and ③ into the appropriate operations indicated in the expression

$$\sigma_s = \sqrt{\frac{③}{③ - 1}}\sqrt{\frac{②}{③} - \left(\frac{①}{③}\right)^2}$$

A solution of this by proper sequencing of operations on a keyboard should not be difficult for anyone at all adept at a calculator.

More important, this is the basis for setting up a simple program in any machine-language or Fortran-language programmable calculator or larger computer. As noted previously, a computer installation generally already has a standard deviation subroutine in the program library, ready to be called by an appropriate command when needed. However for certain desk-top or mini-computers or certain pocket calculators, one may be obliged to compile the program, which is then read onto a cassette tape or magnetic card for handy use. In such cases, the preceding analysis should prove useful.

As an aid to compiling such a program, Table G.1 supplies the step-by-step results in finding the standard deviation (and the mean) for column 1 of Group 1.

Table G.1

	①	②	③		
X	ΣX	X^2	n	σ_s	$\overline{X}$
56.9	56.9	3237.61	1	—	—
58.1	115.0	6613.22	2	0.85	57.50
54.5	169.5	9583.47	3	1.83	56.50
54.0	223.5	12,499.47	4	1.95	55.88
57.1	280.6	15,759.88	5	1.78	56.12
57.8	338.4	19,100.72	6	1.73	56.40
57.3	395.7	22,384.01	7	1.62	56.53
57.9	453.6	25,736.42	8	1.57	56.70
58.1	511.7	29,112.03	9	1.54	56.86
56.8	568.5	32,338.27	10	1.45	56.85

Mean:

$\overline{X}=56.85$ (for the 10 values of X above)

Standard deviation:

$$\sigma_s = \pm 1.45$$

Standard error:

$$\sigma_m = \frac{\sigma_s}{\sqrt{n}} = \pm\frac{1.45}{\sqrt{10}} = \pm 0.46$$

These are the steps in performing the standard deviation computation on a hand calculator which uses reverse Polish notation (RPN) and has a $\Sigma+$ key, such as the HP 80:

1. Key the first value; press $\Sigma+$ key.
2. Continue thus with each value in turn.
   The x-register will successively accumulate and display the values of column ①.
   The y-register will successively accumulate but will not display the values of column ③. To see these at any time, either press $x \rightleftarrows y$ or bring them into view by pressing the $R\downarrow$ (roll-down) key.
   The z-register will successively accumulate but will not display the

values of column ②. To see these at any time, press the *R*↓ key twice. (Before continuing, restore all values to their correct locations.)

3. When the last value has been processed, press $\bar{x}$ to get the mean.
4. Press the $x \rightleftarrows y$ key for the standard deviation.

Any other hand calculator with the standard deviation routine is operated in virtually the same manner.

Appendix H

Problems

Chapter 1

1. If the National Geodetic Survey were to adopt the new value of the inch, assuming that the center of its entire nationwide system of coordinates is in Kansas, by how much would a monument in San Francisco appear to be wrong? (Or would it appear to be wrong?)
2. Could a rubberband serve as a spring balance? Why (not)?
3. Is a digital computer a measuring device? An analog computer?
4. How can the distance from New York to Paris be found by direct measurement? Or, can it?
5. How can the distance to the moon be measured?
6. Tell how one might transfer standards for the following from place to place and utilize them for measurement: time, voltage, resistance, capacitance, mass, length, gravity, temperature, volume, and color.
7. How might one determine his latitude north or south of the equator by measurement with a theodolite or other angle-measuring device?
8. How does the fuel gauge in your auto measure the fuel level in the tank?
9. How does a thermostat sense temperature and translate it into an on/off command to your furnace or radiator?
10. Give one or more advantages of drawing a visual scale on a map over merely noting thereon that the scale is 1:2400 or 1 in. = 200 ft.

Chapter 2

1. What can be done to improve the precision of a single measurement of a distance *AB* (about 35 m long) on a flat surface?
2. What systematic errors must be corrected so that the distance *AB* will approach the true value if a tape is used?
3. What natural errors affect the line of sight when using an engineer's level to establish elevation?
4. How can a carpenter's level be tested to determine trueness?
5. Steel fabrication is very carefully conducted at all temperatures, summer and winter, indoors and outdoors, by using an accurate measuring device, without fear that the steel members will subsequently fail to fit together. What is the nature of the device?
6. What advantage will accrue when measuring the circumference of a steel drum if a steel tape is carefully wrapped around the drum 10 times and the 10× distance is read? Evaluate all the errors of this versus a single wrap-around measurement repeated 10 times.
7. Devise a method of measuring the speed of sound by hammering in a repeating pattern (slowly changing frequency) while a distant observer synchronizes sight and sound. What must be measured?
8. If five stopwatches synchronize perfectly at the end of a 10-min interval, are we assured that
   (*a*) Each runs uniformly?
   (*b*) Each keeps accurate time?
9. How can systematic errors be (*a*) detected? (*b*) eliminated?
10. In Table 2.2, is the "9" (last digit in the mean) justified?
11. How many square feet are there in:

    (*a*) an acre? (*c*) 1.000 acre?

    (*b*) 1.00000 acre? (*d*) 1.00 acre?
12. Do we know the speed of light as 186,000.000 miles per second? What is the value exactly?
13. Is it valid to suggest that the mean in the example of Section 2.11 should be given as 2.14690? Why (not)?
14. Experiment with a polar planimeter to measure the area of each of several regular and irregular figures on a diagram or a map. Calibrate the instrument by use of its accompanying standard-area device, circling it 10 times and dividing the result by 10.

Chapter 3

(Starred problems pertaining to Appendix B are included here but may be omitted without detriment to the continuity.)

1. Which is more valuable to a baseball team, a pitcher with good precision or one with good accuracy?
2. Draw a histogram to depict the following results of 10 pennies tossed together 100 times.

Number of heads	0	1	2	3	4	5	6	7	8	9	10
Frequency	2	1	9	14	16	19	17	13	7	2	0

   Join the center tops of the bars to form a frequency-density curve, and discuss its resemblance to a normal distribution curve. What should be expected if there were 1000 tosses? How do you explain the distribution at the left end?
3. Can we say that any one of the following sets of numbers is a normal distribution? Why (not)?
   (*a*) All the numbers from 1 through 25.
   (*b*) 2.38, 2.37, 2.38, 2.36, 2.36, 2.38, 2.37, 2.36, 2.37, 2.36, 2.36, 2.37, 2.37, 2.38, 2.37, 2.38, 2.37, 2.36, 2.37, 2.38, 2.37, 2.38, 2.37, 2.36.
   (*c*) 143, 140, 144, 141, 144, 142, 141, 142, 140, 145, 143, 143, 142, 143, 142, 143.

4.* In one roll of a single die, what is the probability of rolling a 7? A 5? A 5 or a 2? A 2, a 3, or a 4?

5.* In five rolls of a single die, what is the probability of rolling a 3 once? Twice? Five times? (See Appendix B.)

6.* What is the probability of rolling a 7 with a pair of dice three consecutive times?

7.* Plot a histogram of the sum of four dice shaken and dropped 100 (or 200) times, using along the horizontal axis the values from 4 to 24.

8.* Horse A in the first race has a 1 in 5 chance (probability) of winning; horse B in the second race has a 4 in 3 chance. Using the reasoning of Appendix B, determine the following:

(*a*) What is the probability that, if I bet on each, at least one horse will win?

(*b*) What is the probability that, if I bet on the daily double, both horses will win?

9.* In a handful of 25 coins (8 quarters, 5 dimes, 7 nickels, and 5 pennies), what is the probability of picking from the batch (fairly):

(*a*) A nickel? (*c*) Two nickels (in two picks)?

(*b*) A dime *or* a nickel? (*d*) Three nickels (in three picks)?

10* On a certain test under frigid conditions, 4 out of 10 cars will start up. What prospect (probability) is there that in a new batch of 7, all will start? The prospect that one will start among 7, or 6, or 5, ...?

Hint. The failure rate is 6 in 10, or 0.6; the probability of total failure of all seven is $(0.6)^7 = 0.028$ or 2.8%. This means a 97.% chance that one of seven will start. For the remaining six, the probability of total failure is $(0.6)^6 = 0.047, \ldots$

Chapter 4

1. Of 140 students at camp, the distribution of ages is as given below. Find the standard deviation, the mean, and the standard error of the mean.

17.0– 0	18.0– 8	19.0–5
.1 – 1	.1–11	.1–1
.2– 0	.2– 6	.2–0
.3– 1	.3–10	.3–2
.4– 3	.4– 8	.4–6
.5– 8	.5– 7	.5–0
.6– 5	.6– 6	.6–1
.7– 8	.7– 9	.7–0
.8–10	.8– 5	.8–2
.9–14	.9– 3	.9–0

2. Draw the histogram and try to fit a probability curve to it. Then draw in the σ_s, $2\sigma_s$, and $3\sigma_s$, ordinates.

Hint. Try grouping by class size or class interval in such a way that the histogram and distribution curve look meaningful, rather than using a histogram from the 0.1 class interval, which might be tempting.

3. From the sets of numbers in Problem 3 of Chapter 3, compute for each set a frequency distribution table, and from that the mean, the standard deviation, the standard error, and the best value of the grouping.
4. A deck of playing cards with all face cards removed is shuffled and the top card is turned up. What is the probability that it is either a 5, a 6, or a 7? (Find σ_s, and then what?)
5. If the population of the United States in 1960 was 178,350,000 and that of 1950 was 162,561,000, find that of 1955 by using (*a*) the arithmetic mean, and (*b*) the geometric mean.
6. The following data were obtained from a radar speed detector set up discreetly beside a big city parkway to secure speed information during a free-flowing traffic period. Two different days were used, Tuesday and Wednesday. On the frequency distribution tabulation sheets provided at the end of Appendix F, tally the frequencies, and for each day:
   (*a*) Compute the mean.
   (*b*) Compute the standard deviation.
   (*c*) Compute the standard error.
   (*d*) Plot the normal probability curve.
   (*e*) Plot the normal probability distribution curve.
   (*f*) Plot the data on arithmetic probability paper.
   Compare the behavior of traffic on the two days.

Tuesday speeds (mph)

51	58	50	49	42	54	52	49	43	47
52	50	55	47	52	54	49	42	51	50
45	49	47	53	54	40	53	36	50	49
46	51	47	39	47	47	45	48	39	44
48	45	53	51	50	53	55	42	55	48
55	53	54	54	51	47	40	47	45	54
38	50	55	51	52	50	53	44	44	55
43	51	49	33	39	47	48	54	51	53
54	48	47	55	46	45	59	54	47	46
53	44	53	44	46	54	37	43	56	50

Wednesday speeds (mph)

6	53	46	48	53	58	54	42	51	49
49	30	53	48	57	48	44	49	53	41
57	29	57	34	54	52	45	47	44	50
36	51	50	41	41	39	52	36	53	42
47	47	48	41	49	51	46	59	38	49
54	52	46	50	52	47	40	47	40	49
36	38	54	59	46	52	50	52	42	49
41	52	42	50	44	56	36	42	46	45
47	55	54	48	54	45	35	52	43	55
59	52	45	56	60	38	53	54	54	51

7. Set up a frequency distribution table (as indicated in Appendix F) to calculate the mean and the standard deviation of the values in this sample, as also the standard error of the mean. The values are 3.3, 4.0, 3.7, 3.5, 3.7, 3.9, 3.7, 3.6, 3.9, 3.8, 3.6, 3.5, 3.9, 3.5, 3.6, 3.7, 3.4, 3.6, 3.8 and 3.5 minutes.

8. These are geometry test scores for 16 specially selected scholarship students. Find the standard deviation and the standard mean; draw the histogram with its frequency-density curve; and plot the distribution on arithmetic probability paper (see Appendix F).

Score	f
100	1
99	2
98	5
97	4
96	2
95	2

9. Using the data of Example 4 of Section 4.9 (439 rod readings) calculate the mean, standard deviation, and standard error of the mean by assuming some convenient mean to work with, say 6.500.

Chapter 5

1. The weather reporters frequently say something like this: "We're running 4 degrees below the normal mean for this date. The tempera-

ture is now 47° F; the high was 52°F at 2 P.M. and the low was 35° F at 12 minutes after 7 this morning." Explain the several implications of the statement, such as how the mean is computed, telling the normal mean for the date, and so on.

2. Test the σ_s, $2\sigma_s$, and $3\sigma_s$ values against the data of the second example of Section 4.9, seeking to ascertain whether the proper percentage inclusions are borne out. Discuss any lack of conformity.
3. For Problem 1, Chapter 4, find the median and the mode.
4. In Example 1 Section 5.13, find the median and the mode.
5. If the probable error of a length is ± 0.0764 yd, what is the 3σ error? The 90% error? The "maximum" error?
6. The harmonic mean, not mentioned in the text, is useful for averaging speeds. It is expressed thus:

$$M_{\text{harm}} = \frac{n}{\Sigma(1/x)}$$

If an automobile travels around a one-mile square, the first leg at 40 mhp, the second at 30 mph, the third at 20 mph, and the fourth at 10 mhp, what is its average speed as compared thus? (Let x = speed, n = number of sides.) Compare this with the "average" speed as computed by the arithmetic mean.

7. Using the data of Problem 8, of Chapter 4, find the mean, the median, and the mode for the class. Discuss any significant meaning discovered. Find the standard deviation. What does it say about the class and/or about the students individually.
8. If for a construction project 292 cylinders were tested (28-day, Class A concrete) and an average strength was found to be 4881 psi, with a standard deviation of ± 540 psi, find from an arithmetic probability plot the strength of 99% of the concrete tested (i.e., less than a 1% chance that a test would result in a compressive strength of less than ________ psi). The specifications on the project merely stated a required minimum of 3000 psi. Were they met? Discuss various meanings of "meeting the requirement."

Chapter 6

1. A survey of the ages of 100 of the 150 students in a class was made by one person, and a similar survey of 60 of the 150 was made indepen-

dently by another person:

$$\text{mean } A = 16.87 \pm 0.24 \qquad \text{mean } B = 17.19 \pm 0.31$$

The standard errors of the mean are given.

(*a*) What is the standard deviation in each set?

(*b*) How can it be reasonably established that these two surveys were made of the same group (population)?

2. The following data represent measurements of machine shafts taken from production at random, presumably representing the entire production. What is the average diameter of the shaft, and what is the maximum tolerance that must be established if 95% of the shafts are to be acceptable? What degree of precision does this represent?

2.0001	1.9999	1.9999	2.0001
2.0002	2.0003	2.0000	1.9999
1.9998	2.0001	2.0001	1.9997
1.9999	2.0000	2.0000	2.0000
2.0000	2.0000	1.9998	2.0002

3. Around 1950, by using Geodimeters, Bergstrand computed the velocity of infrared light to be 299,793.1 ± 0.25 km/s and Aslakson independently computed it to be 299,792.4 ± 2.4 km/s. Determine whether these two values are significantly different.

4. Shortly thereafter Aslakson, using other equipment, computed the velocity of radio waves to be 299,794.2 ± 1.4 km/s. Is the velocity of radio waves significantly different from that of light waves?

5. In 1964 NASA published the velocity of light as 299,792.5 ± 0.3 km/s, a still different value from those of Problem 3. Using the three values, which are indeed means of sets of values, calculate the mean, the standard deviation, and the standard error of the mean by treating them as a sample of three values.

6. These are geometry test scores for 24 good students. Calculate the standard deviation and the standard error of this set and test whether this class is significantly different from the class in Problem 8, Chapter 4.

Scores	f
100–98	2
97–95	3
94–92	3
91–89	7
88–86	5
85–83	1
82–80	3

7. Two sets of tests were made, with these results:

Set$A$, $n=12$	*SetB*, $n=16$
$\bar{X}=163.712$	$\bar{X}=163.507$
$\sigma_s=\pm 0.003$	$\sigma_s=\pm 0.006$

Is there one chance in 20 that the two sets are not part of the same population?

Chapter 7

1. A right angle AOB is set off from line OA, and point B is set at a distance of exactly 200 ft. If the angle is sure to within $\pm 01''$, how certain is point B? (Show by enlarged sketch.)
2. Compute the uncertainty (again using a sketch) in the case of Problem 1 if this time the distance OB is 200.000 ± 0.008 ft (i.e., maximum error).
3. Three segments of a line AB are measured several times and the mean of each set is given, along with standard error (σ_m):
   (*a*) 461.812 ± 0.027
   (*b*) 201.003 ± 0.009
   (*c*) 91.161 ± 0.002
   Find the sum and its standard error.
4. The area of a right triangle is to be computed by its base and altitude:

Base	410.817 ± 0.050	($2\sigma_m$ error)
Altitude	101.326 ± 0.025	($2\sigma_m$ error)

Find the area and its $2\sigma_m$ error.

5. A distance AB is measured as 1427.28 ± 0.52 ft along a bearing of N $45°\,00'\,00.0''$ E $\pm 15''$. If the coordinates of A are fixed as N 6792.14 and E 11,792.62, find
   (*a*) The coordinates of B.
   (*b*) The error in northing (latitude) of the course AB.
   (*c*) The error in easting (departure) of the course AB.
   Hint. See Appendix A before fixing upon the square root of 2 for use in this problem.

Chapter 8

1. In a pentagon the angles were measured as shown below (with the maximum error in each given). Find the adjusted angles.
   (*a*) $107°\,10'\,20.0'' \pm 05.0''$
   (*b*) $140°\,27'\,41.7'' \pm 02.5''$
   (*c*) $97°\,50'\,19.3'' \pm 02.5''$
   (*d*) $128°\,23'\,37.5'' \pm 05.0''$
   (*e*) $66°\,08'\,30.0'' \pm 10.0''$
2. Assuming that there is a further set of independent measurements of the complete length AB of Problem 3, Chapter 7, giving 753.989 ± 0.012, find the weighted mean of the length AB.
3. The following data were taken on the dimensions of a triangular plot of ground. What is the best value for the area? What is the 90% error in each side's measurement? Using the 90% errors, what will be the expected error of the area?

Measurement of altitude	Measurement of base
10.001	18.567
10.002	18.566
9.998	18.568
9.999	18.566
10.000	18.566
10.000	18.565
10.001	18.566
9.999	18.567
10.000	18.565
10.010	18.564

4. The developed horsepower in a direct-current motor may be expressed by the following relationship:

$$P_m = \frac{2\pi TN}{33{,}000}$$

where

$$P_m = \text{mechanical power}$$

$$T = \text{torque (lb-ft)}$$

$$N = \text{speed (rpm)}$$

What would be the "expected" power of a single motor and the standard deviation therefrom based on the following experimental data from the motors coming off the assembly line?

Measured torque (lb-ft)	Measured speed (rpm)
62.55	1706
62.57	1704
62.56	1702
62.53	1703
62.55	1704
62.55	1702
62.54	1705
62.56	1703
62.54	1704
62.55	1705
62.54	1704
62.53	1703
62.56	1706
62.55	1704
62.58	1705

5. The following tabulation shows the results of observations made with a one-second theodolite. Each major group of 50 readings was recorded by a single observer, who used these procedures:
   (*a*) The instrument was pointed at a well-defined target; the graduations were matched; and the reading (in seconds) was made, by estimation, to the nearest tenth second.

(*b*) The micrometer wheel was turned to destroy the matching of the graduations, the graduations were then rematched, and a second reading was made. This operation was repeated to a total of 10 readings—the left column in each group.

(*c*) By turning the tangent screw, the instrument was thrown off the target, then repointed on the target, and another series of 10 micrometer readings was made—the second column.

(*d*) Each observer followed this routine for a total of five separate pointings on the target (about 100 ft away from the instrument), and 10 micrometer readings were made for each pointing—the third, fourth, and fifth columns.

Utilizing the "estimated mean" technique, work each column to find the mean of column, standard deviation, and standard error of the mean. For the five columns, then find the weighted mean of all five columns. Compare this result with the mean as determined by treating en masse the 50 observations for each observer.

Group A

56.4	56.0	53.9	51.9	54.5
57.3	56.8	57.6	51.4	57.1
59.7	56.3	58.1	51.8	57.1
58.8	58.1	55.3	52.0	57.5
57.3	57.1	58.4	51.2	56.1
57.3	57.7	57.4	51.0	58.1
56.7	58.0	56.9	53.5	56.8
58.1	58.0	58.5	52.8	57.0
58.3	53.5	59.1	52.7	58.3
57.3	57.4	58.3	52.6	57.0

Group B

57.8	58.0	62.9	60.5	58.2
57.0	58.1	62.0	62.0	58.9
60.0	60.0	61.2	63.0	59.2
55.5	59.6	60.0	59.7	58.2
59.7	60.0	59.5	59.5	58.1
60.0	61.2	58.2	62.2	59.1
69.9	60.0	60.0	63.1	59.3
58.3	59.9	59.0	61.2	60.8
58.8	60.1	59.1	59.2	60.0
57.5	58.9	57.5	62.0	63.7

Group C

58.9	60.0	60.4	61.3	59.8
60.9	60.3	60.9	62.7	60.4
57.7	58.4	60.4	62.7	60.4
54.8	58.8	60.7	62.8	61.1
54.2	59.4	60.4	61.1	59.9
59.8	54.9	61.1	61.7	61.1
57.7	59.8	61.4	61.7	61.1
59.2	60.7	61.1	61.8	60.8
59.8	60.3	60.2	62.4	60.4
57.5	60.1	60.5	61.0	59.9

Group D

51.8	49.6	56.0	53.6	56.0
52.0	52.8	55.9	53.1	55.0
54.0	51.8	55.0	52.2	56.5
53.7	52.8	54.0	55.1	56.0
51.2	52.0	54.9	54.1	55.8
51.8	54.1	54.4	53.0	56.3
52.0	52.0	53.9	54.2	56.1
51.5	54.0	55.3	53.8	55.0
50.8	51.5	53.0	56.0	54.8
50.4	51.0	55.2	56.0	56.8

Group E

55.0	54.9	54.2	55.6	50.1
53.8	55.9	53.0	56.0	50.2
55.7	55.9	53.0	57.0	54.7
57.1	57.4	55.1	57.3	53.2
56.3	54.0	54.2	55.2	52.8
55.9	55.0	52.0	55.4	49.1
56.3	55.3	52.0	54.0	50.1
60.4	54.0	51.5	54.2	49.2
57.3	57.0	52.8	54.2	53.0
57.0	55.3	53.7	55.9	49.2

6. Taking the three values of the speed of light (Problem 5, Chapter 6), find the weighted mean of the three.
7. Find the weighted mean of the three sets of measurements of Section 4.10. Find also the standard error of this new mean (see Section 7.9).

8. Find the weighted mean of the two values (sets A and B) of the Example 1 of Section 6.8. Find also the standard error of this new mean.

Chapter 9

1. Write a specification for laying out a 1000-m distance for an Olympic race that must be accurate to ± 0.2 m. Assume that there is a 50-m-tape available that is correct to ± 0.0005 m under the prevailing temperature, tension, and other handling conditions. Merely ascertain what precision must be used in marking the tape lengths.
2. Calculate the illustrative problem of Appendix C (as suggested in step 4) until a precision of 1 : 10,000 is obtained.
   Hint. Note what change will occur in the e^2 column for any individual change.
   Write the specifications.
3. Calculate the illustrative problem of Appendix C by using the E_t as cumulative (as suggested in step 2*b*.) Write the specifications.
4. Calculate the illustrative problem of Appendix C by holding the tape on the ground (e.g., assume smooth pavement). Assume that a hand level is used to measure the slope between tape ends to ± 1 ft.
   Hint. Will e_g then be cumulative? What becomes of e_s? How would you rewrite the specification for marking?
5. Calculate the problem by using a 50-ft steel tape.
6. Several elevation determinations were made on a questionable foundation at La Guardia Airport during the autumn and several more during the spring, always working from a firm bench mark 0.8 mile away. From the data, calculate the elevation in the fall, with its various sigma errors, and again in the spring, to see whether there has been definite settlement. (Assume that the "closure" values of Appendix D are in this case equivalently the $2\sigma_s$ error of each level run.)

Autumn			Spring		
Run Number	Elevation	Order	Run Number	Elevation	Order
1	8.2462	Third	7	8.2479	Second
2	8.2478	First	8	8.2452	Third
3	8.2467	Second	9	8.2467	First
4	8.2481	First	10	8.2473	Second
5	8.2480	First	11	8.2479	First
6	8.2469	Second	12	8.2475	Second

7. A research experiment resulted in four sets of the result ($n = 25$ in each set), and it is desired to find the weighted mean of the four results, with the standard error of the result.

Set	Mean (mm)	σ_s (mm)
A	124.390	±1.589
B	124.252	±1.540
C	124.728	±1.937
D	124.936	±1.752

Chapter 10

1. A lighted target 10,000 ft distant is observed with a theodolite to within one-tenth second of angle (by use of a repetition technique). The distance to the point is measured by a radar-frequency distance-measuring device to within 0.012 ft. Is there consistency between angular and distance techniques? If not, ascertain what should be the proper angular error value.
2. If point *B* in the preceding problem is located with a 50% assurance within a circle of 0.01 ft radius, what is the radius of the circle for 90% assurance?
3. With present technology it is feasible to measure from Florida to California (approximately 3000 miles) and be 50% sure of distance within ±140 ft. If a theodolite were used to measure the angles between adjoining segments of this line (assuming 300 angles), what angular error would be the maximum permissible?
4. Assuming that a range finder can give distance of 3000 ft within a 90% accuracy of ±200 ft, what angular accuracy would be sufficient? What if the distance accuracy were ±100 ft? ±10 ft?

Index

Accidental error, 9, 13, 17, 25
Accuracy, 15, 26, 109
  positional, 109
Adjustment, 85, 88
Ampere, 3
Angular error, 108
Arithmetic mean, 8, 14, 24, 46, 84
Arithmetic probability, 60, 64, 72, 154, 158
Aslaskon, 172
Astronomic position, 148
Average, 8, 24, 46, 80

Bergstrand, 172
Best value, 9, 14
Blunder, 9

Calculator, 115, 159, 161
Candela, 3
Central tendency, 46
Certainty, 50, 73
Class interval, 41
Coefficient, of variance, 49
  of variation, 50
Computers, 159
Confidence, 21, 53, 55
Control systems, 111, 135
Correction, 12, 88
Curve-plotting, 127, 153, 154

Deflection of vertical, 145, 148, 149
Deviation, 48, 78
Discrepancy, 9, 13
Doubtful figures, 113

Earth, 144
Ellipsoid, 144
Error, 7, 23, 140
  accidental, 9, 13, 17, 25
  circular, 107
  conversion, 55
  cumulative, 10
  elliptical, 107
  instrumental, 11
  maximum, 54, 68, 73, 95
  natural, 11
  personal, 11
  probable, 52, 54
  propagation, 72
  random, 23, 72
  relative, 110
  sigma, 50
  standard, 30, 39, 53, 66, 69, 78, 87, 90, 140, 160
  summary table, 54
  systematic, 9, 10
  two-dimensional, 103
  and weights, 80, 87
Estimation, 8
Exact figures, 113, 116

Fortran, 159
Frequency, 3, 46
Frequency density curve, 18, 39
Frequency distribution, 32, 151, 157

Geodetic, 135, 143
Geodimeter, 172
Geoid, 144
Gravity, 3, 141, 144, 147

Histogram, 17, 39, 119, 129

Inch, 2, 4
Indexes, 83

Judgment, 82

Least square principle, 22, 24
Length, 2

Mass, 3
Maximum error, 54, 68, 73, 95
Mean, 14, 46
  arithmetic, 8, 14, 24, 46, 84
  assumed, 34, 36
  weighted, 80, 84
Mean deviation, 48, 78
Mean square error, 27
Mean variation, 78
Measurement, 1
  direct, 6
  error, 7
  indirect, 6
  rejection of, 68
  repeated, 67
Median, 46
Meter, 2
Mistake, 9
Mode, 47
Most probable value, 57

NASA, 172
National Bureau of Standards, 3
National Ocean Survey, NOAA, 135
Normal distribution curve, 19, 45, 54, 129, 151
Normal probability curve, 20, 55, 59, 118, 127

Poisson distribution, 20
Population, 21, 29, 32, 53
Position accuracy, 109
Precision, 15, 26, 66
Precision ratio, 108, 132
Probability, 23, 59, 104, 118
  circle, 105
  curve, 20
    arithmetic, 60, 64, 72, 154, 158
    density, 41, 120
    distribution, 151
    normal, 10, 55, 59, 118, 127
    skewed, 22, 47
  ellipse, 104
  theory, 27
Probable error, 52, 54
Probable value, 24
Procedure, 96, 110, 130
Propagation of errors, 72

Random error, 23, 72
Random occurrence, 20, 23
Range, 48
Readings, 6, 7
Rejecting measurements, 68
Reliability, 15, 46, 66
Repeated measurement, 67
Residuals, 14, 23, 25
Root mean square error, 30
Round-off, 115

Sample, 28, 67, 71, 94, 130
Sample size, 21, 24, 28, 101
Satellites, 148
Scale, 8, 165
Scatter, 48
Sigma error, 30, 50, 57
Significance, 69, 71
Significant difference, 71
Significant figures, 113
Skew, 22, 47
Specifications, 100, 130, 140
Spheroid, 144
Standard, 2, 135
Standard deviation, 27, 29, 34, 49, 53, 66, 94, 159
  of population, 29
Standard error, 30, 39, 53, 66, 69, 78, 87, 90, 140, 160
  of difference, 70
Standard procedure, 96, 100

Statistical measurement, 53
Systematic error, 10, 140

Taping errors, 130
Taping specifications, 130
Three-sigma error, 52, 54
Time, 3
Temperature, 3
Tolerance, 95
True value, 9, 31, 68

Two-sigma error, 51, 54

Undulation, 144

Variation, 16, 20, 50, 144

Weight, 80, 81, 84
Weighted mean, 80, 84

Zero, 114